Collins
World Atlas

Settlements

Population	National capital	Administrative capital	Other city or town
over 10 million	BEIJING ✦	Karachi ◉	New York ◉
5 million to 10 million	LONDON ✦	Tianjin ◉	Santos ◉
1 million to 5 million	KĀBUL ✦	Sydney ◉	Kaohsiung ◉
500 000 to 1 million	BANGUI ✦	Trujillo ◉	Amritsar ◉
100 000 to 500 000	WELLINGTON ✦	Mansa ◉	Apucarana ◉
50 000 to 100 000	PORT OF SPAIN ✦	Potenza ◉	Arecibo ◉
10 000 to 50 000	MALABO ✦	Chinhoyi ○	Ceres ○
under 10 000	VALLETTA ✦	Ati ○	Venta ○

Built-up area

Boundaries

- International boundary
- Disputed international boundary or alignment unconfirmed
- Disputed territory boundary
- Administrative boundary
- Ceasefire line
- UN Buffer zone

Miscellaneous

- National park
- Reserve or Regional park
- Site of specific interest
- Wall

Land and sea features

- Desert
- Oasis
- Lava field
- 1234 Volcano height in metres
- Marsh
- Ice cap or Glacier
- Escarpment
- Coral reef
- 1234 Pass height in metres

Lakes and rivers

- Lake
- Impermanent lake
- Salt lake or lagoon
- Impermanent salt lake
- Dry salt lake or salt pan
- 123 Lake height surface height above sea level, in metres
- River
- Impermanent river or watercourse
- Waterfall
- Dam
- Barrage

Relief

Contour intervals and layer colours

Height

metres		feet
5000		16404
3000		9843
2000		6562
1000		3281
500		1640
200		656
0		0
below sea level		
0		0
200		656
2000		6562
4000		13124
6000		19686

Depth

- 1234 ▲ Summit height in metres
- -123 Spot height height in metres
- 123 Ocean deep depth in metres
- 2000 Ice surface elevation above sea level (in metres)

Transport

- Motorway (tunnel; under construction)
- Main road (tunnel; under construction)
- Secondary road (tunnel; under construction)
- Track
- Main railway (tunnel; under construction)
- Secondary railway (tunnel; under construction)
- Other railway (tunnel; under construction)
- Canal
- ✈ Main airport
- ✈ Regional airport

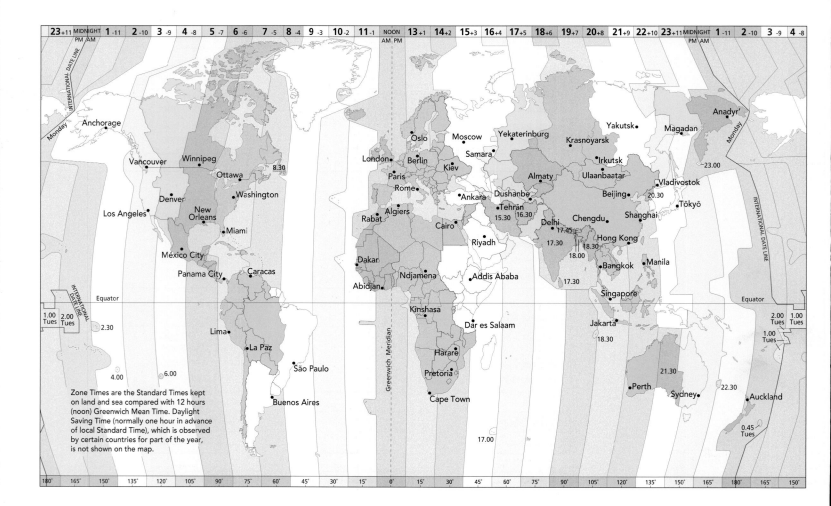

Zone Times are the Standard Times kept on land and sea compared with 12 hours (noon) Greenwich Mean Time. Daylight Saving Time (normally one hour in advance of local Standard Time), which is observed by certain countries for part of the year, is not shown on the map.

Map Symbols and Time Zones

Europe

Europe		Area sq km	Area sq miles	Population	Capital	Languages	Religions	Currency	Internet Link
ALBANIA		28 748	11 100	3 173 000	Tirana	Albanian, Greek	Sunni Muslim, Albanian Orthodox, Roman Catholic	Lek	www.km.gov.al
ANDORRA		465	180	79 000	Andorra la Vella	Catalan, Spanish, French	Roman Catholic	Euro	www.govern.ad
AUSTRIA		83 855	32 377	8 495 000	Vienna	German, Croatian, Turkish	Roman Catholic, Protestant	Euro	www.bundeskanzleramt.at
BELARUS		207 600	80 155	9 357 000	Minsk	Belarusian, Russian	Belarusian Orthodox, Roman Catholic	Belarusian rouble	www.belarus.by
BELGIUM		30 520	11 784	11 104 000	Brussels	Dutch (Flemish), French (Walloon), German	Roman Catholic, Protestant	Euro	www.belgium.be
BOSNIA AND HERZEGOVINA		51 130	19 741	3 829 000	Sarajevo	Bosnian, Serbian, Croatian	Sunni Muslim, Serbian Orthodox, Roman Catholic, Protestant	Convertible mark	www.fbihvlada.gov.ba
BULGARIA		110 994	42 855	7 223 000	Sofia	Bulgarian, Turkish, Romany, Macedonian	Bulgarian Orthodox, Sunni Muslim	Lev	www.government.bg
CROATIA		56 538	21 829	4 290 000	Zagreb	Croatian, Serbian	Roman Catholic, Serbian Orthodox, Sunni Muslim	Kuna	www.vlada.hr
CZECHIA (CZECH REPUBLIC)		78 864	30 450	10 702 000	Prague	Czech, Moravian, Slovak	Roman Catholic, Protestant	Koruna	www.czech.cz
DENMARK		43 075	16 631	5 619 000	Copenhagen	Danish	Protestant	Danish krone	www.denmark.dk
ESTONIA		45 200	17 452	1 287 000	Tallinn	Estonian, Russian	Protestant, Estonian and Russian Orthodox	Euro	https://valitsus.ee
FINLAND		338 145	130 559	5 426 000	Helsinki	Finnish, Swedish, Sami languages	Protestant, Greek Orthodox	Euro	http://valtioneuvosto.fi
FRANCE		543 965	210 026	64 291 000	Paris	French, German dialects, Italian, Arabic, Breton	Roman Catholic, Protestant, Sunni Muslim	Euro	www.premier-ministre.gouv.fr
GERMANY		357 022	137 849	82 727 000	Berlin	German, Turkish	Protestant, Roman Catholic	Euro	www.bundesregierung.de
GREECE		131 957	50 949	11 128 000	Athens	Greek	Greek Orthodox, Sunni Muslim	Euro	www.primeminister.gr
HUNGARY		93 030	35 919	9 955 000	Budapest	Hungarian	Roman Catholic, Protestant	Forint	https://magyarorszag.hu
ICELAND		102 820	39 699	330 000	Reykjavík	Icelandic	Protestant	Icelandic króna	www.iceland.is
IRELAND		70 282	27 136	4 627 000	Dublin	English, Irish	Roman Catholic, Protestant	Euro	www.gov.ie
ITALY		301 245	116 311	60 990 000	Rome	Italian	Roman Catholic	Euro	www.governo.it
KOSOVO		10 908	4 212	1 815 606	Pristina (Prishtinë)	Albanian, Serbian	Sunni Muslim, Serbian Orthodox	Euro	www.rks-gov.net
LATVIA		64 589	24 938	2 050 000	Rīga	Latvian, Russian	Protestant, Roman Catholic, Russian Orthodox	Euro	www.saeima.lv
LIECHTENSTEIN		160	62	37 000	Vaduz	German	Roman Catholic, Protestant	Swiss franc	www.liechtenstein.li
LITHUANIA		65 200	25 174	3 017 000	Vilnius	Lithuanian, Russian, Polish	Roman Catholic, Russian Orthodox	Euro	www.lrv.lt
LUXEMBOURG		2 586	998	530 000	Luxembourg	Letzeburgish, German, French	Roman Catholic	Euro	www.gouvernement.lu
MALTA		316	122	429 000	Valletta	Maltese, English	Roman Catholic	Euro	www.gov.mt
MOLDOVA		33 700	13 012	3 487 000	Chișinău	Romanian, Ukrainian, Gagauz, Russian	Romanian Orthodox, Russian Orthodox	Moldovan leu	www.moldova.md
MONACO		2	1	38 000	Monaco-Ville	French, Monegasque, Italian	Roman Catholic	Euro	www.monaco.gouv.mc
MONTENEGRO		13 812	5 333	621 000	Podgorica	Serbian (Montenegrin), Albanian	Montenegrin Orthodox, Sunni Muslim	Euro	www.gov.me
NETHERLANDS		41 526	16 033	16 759 000	Amsterdam/The Hague	Dutch, Frisian	Roman Catholic, Protestant, Sunni Muslim	Euro	www.overheid.nl
NORTH MACEDONIA		25 713	9 928	2 107 000	Skopje	Macedonian, Albanian, Turkish	Macedonian Orthodox, Sunni Muslim	Macedonian denar	www.vlada.mk
NORWAY		323 878	125 050	5 043 000	Oslo	Norwegian, Sami languages	Protestant, Roman Catholic	Norwegian krone	www.norge.no
POLAND		312 683	120 728	38 217 000	Warsaw	Polish, German	Roman Catholic, Polish Orthodox	Złoty	www.poland.gov.pl
PORTUGAL		88 940	34 340	10 608 000	Lisbon	Portuguese	Roman Catholic, Protestant	Euro	www.portugal.gov.pt
ROMANIA		237 500	91 699	21 699 000	Bucharest	Romanian, Hungarian	Romanian Orthodox, Protestant, Roman Catholic	Romanian leu	www.guv.ro
RUSSIA		17 075 400	6 592 849	142 834 000	Moscow	Russian, Tatar, Ukrainian, other local languages	Russian Orthodox, Sunni Muslim, Protestant	Russian rouble	www.gov.ru
SAN MARINO		61	24	31 000	San Marino	Italian	Roman Catholic	Euro	www.consigliograndeegenerale.sm
SERBIA		77 453	29 904	7 181 505	Belgrade	Serbian, Hungarian	Serbian Orthodox, Roman Catholic, Sunni Muslim	Serbian dinar	www.srbija.gov.rs
SLOVAKIA		49 035	18 933	5 450 000	Bratislava	Slovak, Hungarian, Czech	Roman Catholic, Protestant, Orthodox	Euro	www.government.gov.sk
SLOVENIA		20 251	7 819	2 072 000	Ljubljana	Slovene, Croatian, Serbian	Roman Catholic, Protestant	Euro	www.gov.si
SPAIN		504 782	194 897	46 927 000	Madrid	Spanish (Castilian), Catalan, Galician, Basque	Roman Catholic	Euro	www.lamoncloa.gob.es
SWEDEN		449 964	173 732	9 571 000	Stockholm	Swedish, Sami languages	Protestant, Roman Catholic	Swedish krona	www.sweden.se
SWITZERLAND		41 293	15 943	8 078 000	Bern	German, French, Italian, Romansch	Roman Catholic, Protestant	Swiss franc	www.eda.admin.ch/aboutswitzerland
UKRAINE		603 700	233 090	45 239 000	Kiev	Ukrainian, Russian	Ukrainian Orthodox, Ukrainian Catholic, Roman Catholic	Hryvnia	www.kmu.gov.ua
UNITED KINGDOM		243 609	94 058	63 136 000	London	English, Welsh, Gaelic	Protestant, Roman Catholic, Muslim	Pound sterling	www.gov.uk
VATICAN CITY		0.5	0.2	800	Vatican City	Italian	Roman Catholic	Euro	www.vaticanstate.va

Asia

Asia		Area sq km	Area sq miles	Population	Capital	Languages	Religions	Currency	Internet Link
AFGHANISTAN		652 225	251 825	30 552 000	Kābul	Dari, Pashto (Pashtu), Uzbek, Turkmen	Sunni Muslim, Shi'a Muslim	Afghani	www.president.gov.af
ARMENIA		29 800	11 506	2 977 000	Yerevan	Armenian, Kurdish	Armenian Orthodox	Dram	www.gov.am
AZERBAIJAN		86 600	33 436	9 413 000	Baku	Azeri, Armenian, Russian, Lezgian	Shi'a Muslim, Sunni Muslim, Russian and Armenian Orthodox	Azerbaijani manat	www.president.az
BAHRAIN		691	267	1 332 000	Manama	Arabic, English	Shi'a Muslim, Sunni Muslim, Christian	Bahrain dinar	www.bahrain.bh
BANGLADESH		143 998	55 598	156 595 000	Dhaka	Bengali, English	Sunni Muslim, Hindu	Taka	www.bangladesh.gov.bd
BHUTAN		46 620	18 000	754 000	Thimphu	Dzongkha, Nepali, Assamese	Buddhist, Hindu	Ngultrum, Indian rupee	www.bhutan.gov.bt
BRUNEI		5 765	2 226	418 000	Bandar Seri Begawan	Malay, English, Chinese	Sunni Muslim, Buddhist, Christian	Bruneian dollar	www.pmo.gov.bn
CAMBODIA		181 035	69 884	15 135 000	Phnom Penh	Khmer, Vietnamese	Buddhist, Roman Catholic, Sunni Muslim	Riel	www.cambodia.gov.kh
CHINA		9 606 802	3 709 186	1 369 993 000	Beijing	Mandarin (Putonghua), Wu, Cantonese, Hsiang, regional languages	Confucian, Taoist, Buddhist, Christian, Sunni Muslim	Yuan, HK dollar*, Macau pataca	www.gov.cn
CYPRUS		9 251	3 572	1 141 000	Nicosia	Greek, Turkish, English	Greek Orthodox, Sunni Muslim	Euro	www.cyprus.gov.cy
EAST TIMOR (TIMOR-LESTE)		14 874	5 743	1 133 000	Dili	Portuguese, Tetun, English	Roman Catholic	United States dollar	http://timor-leste.gov.tl
GEORGIA		69 700	26 911	4 341 000	Tbilisi	Georgian, Russian, Armenian, Azeri, Ossetian, Abkhaz	Georgian Orthodox, Russian Orthodox, Sunni Muslim	Lari	www.parliament.ge
INDIA		3 166 620	1 222 632	1 252 140 000	New Delhi	Hindi, English, many regional languages	Hindu, Sunni Muslim, Shi'a Muslim, Sikh, Christian	Indian rupee	www.india.gov.in
INDONESIA		1 919 445	741 102	249 866 000	Jakarta	Indonesian, other local languages	Sunni Muslim, Protestant, Roman Catholic, Hindu, Buddhist	Rupiah	www.indonesia.go.id
IRAN		1 648 000	636 296	77 447 000	Tehrān	Farsi, Azeri, Kurdish, regional languages	Shi'a Muslim, Sunni Muslim	Iranian rial	www.president.ir
IRAQ		438 317	169 235	33 765 000	Baghdād	Arabic, Kurdish, Turkmen	Shi'a Muslim, Sunni Muslim, Christian	Iraqi dinar	www.cabinet.iq
ISRAEL		22 072	8 522	7 733 000	Jerusalem (Yerushalayim) (El Quds)**	Hebrew, Arabic	Jewish, Sunni Muslim, Christian, Druze	Shekel	www.knesset.gov.il
JAPAN		377 727	145 841	127 144 000	Tōkyō	Japanese	Shintoist, Buddhist, Christian	Yen	www.japan.go.jp
JORDAN		89 206	34 443	7 274 000	'Ammān	Arabic	Sunni Muslim, Christian	Jordanian dinar	www.jordan.gov.jo
KAZAKHSTAN		2 717 300	1 049 155	16 441 000	Astana	Kazakh, Russian, Ukrainian, German, Uzbek, Tatar	Sunni Muslim, Russian Orthodox, Protestant	Tenge	www.government.kz
KUWAIT		17 818	6 880	3 369 000	Kuwait	Arabic	Sunni Muslim, Shi'a Muslim, Christian, Hindu	Kuwaiti dinar	www.e.gov.kw
KYRGYZSTAN		198 500	76 641	5 548 000	Bishkek	Kyrgyz, Russian, Uzbek	Sunni Muslim, Russian Orthodox	Kyrgyz som	www.gov.kg
LAOS		236 800	91 429	6 770 000	Vientiane	Lao, other local languages	Buddhist, traditional beliefs	Kip	www.na.gov.la
LEBANON		10 452	4 036	4 822 000	Beirut	Arabic, Armenian, French	Shi'a Muslim, Sunni Muslim, Christian	Lebanese pound	www.presidency.gov.lb
MALAYSIA		332 965	128 559	29 717 000	Kuala Lumpur/Putrajaya	Malay, English, Chinese, Tamil, other local languages	Sunni Muslim, Buddhist, Hindu, Christian, traditional beliefs	Ringgit	www.malaysia.gov.my

**De facto capital. Disputed   *Hong Kong dollar

National Statistics   (3)

Asia continued		Area sq km	Area sq miles	Population	Capital	Languages	Religions	Currency	Internet Link
MALDIVES		298	115	345 000	Male	Divehi (Maldivian)	Sunni Muslim	Rufiyaa	www.presidencymaldives.gov.mv
MONGOLIA		1 565 000	604 250	2 839 000	Ulan Bator	Khalka (Mongolian), Kazakh, other local languages	Buddhist, Sunni Muslim	Tugrik (tögrög)	www.pmis.gov.mn
MYANMAR (BURMA)		676 577	261 228	53 259 000	Nay Pyi Taw	Burmese, Shan, Karen, other local languages	Buddhist, Christian, Sunni Muslim	Kyat	www.president-office.gov.mm
NEPAL		147 181	56 827	27 797 000	Kathmandu	Nepali, Maithili, Bhojpuri, English, other local languages	Hindu, Buddhist, Sunni Muslim	Nepalese rupee	www.nepalgov.gov.np
NORTH KOREA		120 538	46 540	24 895 000	P'yŏngyang	Korean	Traditional beliefs, Chondoist, Buddhist	North Korean won	www.korea-dpr.com
OMAN		309 500	119 499	3 632 000	Muscat	Arabic, Baluchi, Indian languages	Ibadhi Muslim, Sunni Muslim	Omani riyal	www.oman.om
PAKISTAN		881 888	340 497	182 143 000	Islamabad	Urdu, Punjabi, Sindhi, Pashto (Pashtu), English, Balochi	Sunni Muslim, Shi'a Muslim, Christian, Hindu	Pakistani rupee	www.pakistan.gov.pk
PALAU		497	192	21 000	Ngerulmud	Palauan, English	Roman Catholic, Protestant, traditional beliefs	United States dollar	http://palaugov.org
PHILIPPINES		300 000	115 831	98 394 000	Manila	English, Filipino, Tagalog, Cebuano, other local languages	Roman Catholic, Protestant, Sunni Muslim, Aglipayan	Philippine peso	http://president.gov.ph/
QATAR		11 437	4 416	2 169 000	Doha	Arabic	Sunni Muslim	Qatari riyal	www.gov.qa
RUSSIA		17 075 400	6 592 849	142 834 000	Moscow	Russian, Tatar, Ukrainian, other local languages	Russian Orthodox, Sunni Muslim, Protestant	Russian rouble	www.gov.ru
SAUDI ARABIA		2 200 000	849 425	28 829 000	Riyadh	Arabic	Sunni Muslim, Shi'a Muslim	Saudi Arabian riyal	www.saudi.gov.sa
SINGAPORE		639	247	5 412 000	Singapore	Chinese, English, Malay, Tamil	Buddhist, Taoist, Sunni Muslim, Christian, Hindu	Singapore dollar	www.gov.sg
SOUTH KOREA		99 274	38 330	49 263 000	Seoul	Korean	Buddhist, Protestant, Roman Catholic	South Korean won	www.korea.net
SRI LANKA		65 610	25 332	21 273 000	Sri Jayewardenepura Kotte	Sinhalese, Tamil, English	Buddhist, Hindu, Sunni Muslim, Roman Catholic	Sri Lankan rupee	www.priu.gov.lk
SYRIA		184 026	71 052	21 898 000	Damascus	Arabic, Kurdish, Armenian	Sunni Muslim, Shi'a Muslim, Christian	Syrian pound	http://parliament.sy
TAIWAN		36 179	13 969	23 344 000	Taipei (Taibei)	Mandarin (Putonghua), Min, Hakka, other local languages	Buddhist, Taoist, Confucian, Christian	New Taiwan dollar	www.gov.tw
TAJIKISTAN		143 100	55 251	8 208 000	Dushanbe	Tajik, Uzbek, Russian	Sunni Muslim	Somoni	www.prezident.tj
THAILAND		513 115	198 115	67 011 000	Bangkok	Thai, Lao, Chinese, Malay, Mon-Khmer languages	Buddhist, Sunni Muslim	Baht	www.thaigov.go.th
TURKEY		779 452	300 948	74 933 000	Ankara	Turkish, Kurdish	Sunni Muslim, Shi'a Muslim	Lira	www.tccb.gov.tr
TURKMENISTAN		488 100	188 456	5 240 000	Ashgabat (Aşgabat)	Turkmen, Uzbek, Russian	Sunni Muslim, Russian Orthodox	Turkmen manat	www.turkmenistan.gov.tm
UNITED ARAB EMIRATES		77 700	30 000	9 346 000	Abu Dhabi (Abū ẓaby)	Arabic, English	Sunni Muslim, Shi'a Muslim	United Arab Emirates dirham	www.government.ae
UZBEKISTAN		447 400	172 742	28 934 000	Toshkent (Tashkent)	Uzbek, Russian, Tajik, Kazakh	Sunni Muslim, Russian Orthodox	Uzbek som	www.gov.uz
VIETNAM		329 565	127 246	91 680 000	Ha Nôi (Hanoi)	Vietnamese, Thai, Khmer, Chinese, other local languages	Buddhist, Taoist, Roman Catholic, Cao Dai, Hoa Hao	Dong	www.na.gov.vn
YEMEN		527 968	203 850	24 407 000	Şan'ā'	Arabic	Sunni Muslim, Shi'a Muslim	Yemeni rial	www.yemen-nic.info

Africa		Area sq km	Area sq miles	Population	Capital	Languages	Religions	Currency	Internet Link
ALGERIA		2 381 741	919 595	39 208 000	Algiers (Alger)	Arabic, French, Berber	Sunni Muslim	Algerian dinar	www.el-mouradia.dz
ANGOLA		1 246 700	481 354	21 472 000	Luanda	Portuguese, Bantu, other local languages	Roman Catholic, Protestant, traditional beliefs	Kwanza	www.governo.gov.ao
BENIN		112 620	43 483	10 323 000	Porto-Novo	French, Fon, Yoruba, Adja, other local languages	Traditional beliefs, Roman Catholic, Sunni Muslim	CFA franc*	www.gouv.bj
BOTSWANA		581 370	224 468	2 021 000	Gaborone	English, Setswana, Shona, other local languages	Traditional beliefs, Protestant, Roman Catholic	Pula	www.gov.bw
BURKINA FASO		274 200	105 869	16 935 000	Ouagadougou	French, Moore (Mossi), Fulani, other local languages	Sunni Muslim, traditional beliefs, Roman Catholic	CFA franc*	www.gouvernement.gov.bf
BURUNDI		27 835	10 747	10 163 000	Bujumbura	Kirundi (Hutu, Tutsi), French	Roman Catholic, traditional beliefs, Protestant	Burundian franc	www.burundi-gov.bi
CAMEROON		475 442	183 569	22 254 000	Yaoundé	French, English, Fang, Bamileke, other local languages	Roman Catholic, traditional beliefs, Sunni Muslim, Protestant	CFA franc*	www.spm.gov.cm
CAPE VERDE (CABO VERDE)		4 033	1 557	499 000	Praia	Portuguese, creole	Roman Catholic, Protestant	Cape Verdean escudo	www.governo.cv
CENTRAL AFRICAN REPUBLIC		622 436	240 324	4 616 000	Bangui	French, Sango, Banda, Baya, other local languages	Protestant, Roman Catholic, traditional beliefs, Sunni Muslim	CFA franc*	www.centrafricaine.info
CHAD		1 284 000	495 755	12 825 000	Ndjamena	Arabic, French, Sara, other local languages	Sunni Muslim, Roman Catholic, Protestant, traditional beliefs	CFA franc*	www.presidencetchad.org
COMOROS		1 862	719	735 000	Moroni	Shikomor (Comorian), French, Arabic	Sunni Muslim, Roman Catholic	Comorian franc	www.beit-salam.km
CONGO		342 000	132 047	4 448 000	Brazzaville	French, Kongo, Monokutuba, other local languages	Roman Catholic, Protestant, traditional beliefs, Sunni Muslim	CFA franc*	www.presidence.cg
CONGO, DEM. REP. OF THE		2 345 410	905 568	67 514 000	Kinshasa	French, Lingala, Swahili, Kongo, other local languages	Christian, Sunni Muslim	Congolese franc	www.president-rdc.cd
CÔTE D'IVOIRE (IVORY COAST)		322 463	124 504	20 316 000	Yamoussoukro	French, creole, Akan, other local languages	Sunni Muslim, Roman Catholic, traditional beliefs, Protestant	CFA franc*	www.gouv.ci
DJIBOUTI		23 200	8 958	873 000	Djibouti	Somali, Afar, French, Arabic	Sunni Muslim, Christian	Djiboutian franc	www.presidence.dj
EGYPT		1 001 450	386 660	82 056 000	Cairo (Al Qāhirah)	Arabic	Sunni Muslim, Coptic Christian	Egyptian pound	www.egypt.gov.eg
EQUATORIAL GUINEA		28 051	10 831	757 000	Malabo	Spanish, French, Fang	Roman Catholic, traditional beliefs	CFA franc*	www.guineaecuatorialpress.com
ERITREA		117 400	45 328	6 333 000	Asmara	Tigrinya, Tigre	Sunni Muslim, Coptic Christian	Nakfa	www.shabait.com
ESWATINI (SWAZILAND)		17 364	6 704	1 250 000	Mbabane/Lobamba	Swazi, English	Christian, traditional beliefs	Lilangeni, South African rand	www.gov.sz
ETHIOPIA		1 133 880	437 794	94 101 000	Addis Ababa	Oromo, Amharic, Tigrinya, other local languages	Ethiopian Orthodox, Sunni Muslim, traditional beliefs	Birr	www.ethiopia.gov.et
GABON		267 667	103 347	1 672 000	Libreville	French, Fang, other local languages	Roman Catholic, Protestant, traditional beliefs	CFA franc*	www.legabon.org
THE GAMBIA		11 295	4 361	1 849 000	Banjul	English, Malinke, Fulani, Wolof	Sunni Muslim, Protestant	Dalasi	www.assembly.gov.gm
GHANA		238 537	92 100	25 905 000	Accra	English, Hausa, Akan, other local languages	Christian, Sunni Muslim, traditional beliefs	Cedi	www.ghana.gov.gh
GUINEA		245 857	94 926	11 745 000	Conakry	French, Fulani, Malinke, other local languages	Sunni Muslim, traditional beliefs, Christian	Guinean franc	www.assemblee.gov.gn
GUINEA-BISSAU		36 125	13 948	1 704 000	Bissau	Portuguese, crioulo, other local languages	Traditional beliefs, Sunni Muslim, Christian	CFA franc*	www.guinebissaurepublic.com
KENYA		582 646	224 961	44 354 000	Nairobi	Swahili, English, other local languages	Christian, traditional beliefs	Kenyan shilling	www.president.go.ke
LESOTHO		30 355	11 720	2 074 000	Maseru	Sesotho, English, Zulu	Christian, traditional beliefs	Loti, S. African rand	www.gov.ls
LIBERIA		111 369	43 000	4 294 000	Monrovia	English, creole, other local languages	Traditional beliefs, Christian, Sunni Muslim	Liberian dollar	www.emansion.gov.lr
LIBYA		1 759 540	679 362	6 202 000	Tripoli	Arabic, Berber	Sunni Muslim	Libyan dinar	www.libyanmission-un.org
MADAGASCAR		587 041	226 658	22 925 000	Antananarivo	Malagasy, French	Traditional beliefs, Christian, Sunni Muslim	Ariary	www.madagascar.gov.mg
MALAWI		118 484	45 747	16 363 000	Lilongwe	Chichewa, English, other local languages	Christian, traditional beliefs, Sunni Muslim	Malawian kwacha	www.malawi.gov.mw
MALI		1 240 140	478 821	15 302 000	Bamako	French, Bambara, other local languages	Sunni Muslim, traditional beliefs, Christian	CFA franc*	www.primature.gov.ml
MAURITANIA		1 030 700	397 955	3 890 000	Nouakchott	Arabic, French, other local languages	Sunni Muslim	Ouguiya	www.mauritania.mr
MAURITIUS		2 040	788	1 244 000	Port Louis	English, creole, Hindi, Bhojpurī, French	Hindu, Roman Catholic, Sunni Muslim	Mauritian rupee	www.gov.mu
MOROCCO		446 550	172 414	33 008 000	Rabat	Arabic, Berber, French	Sunni Muslim	Moroccan dirham	www.maroc.ma
MOZAMBIQUE		799 380	308 642	25 834 000	Maputo	Portuguese, Makua, Tsonga, other local languages	Traditional beliefs, Roman Catholic, Sunni Muslim	Metical	www.portaldogoverno.gov.mz
NAMIBIA		824 292	318 261	2 303 000	Windhoek	English, Afrikaans, German, Ovambo, other local languages	Protestant, Roman Catholic	Namibian dollar	www.gov.na
NIGER		1 267 000	489 191	17 831 000	Niamey	French, Hausa, Fulani, other local languages	Sunni Muslim, traditional beliefs	CFA franc*	www.presidence.ne
NIGERIA		923 768	356 669	173 615 000	Abuja	English, Hausa, Yoruba, Ibo, Fulani, other local languages	Sunni Muslim, Christian, traditional beliefs	Naira	www.nigeria.gov.ng
RWANDA		26 338	10 169	11 777 000	Kigali	Kinyarwanda, French, English	Roman Catholic, traditional beliefs, Protestant	Rwandan franc	www.gov.rw
SÃO TOMÉ AND PRÍNCIPE		964	372	193 000	São Tomé	Portuguese, creole	Roman Catholic, Protestant	Dobra	www.gov.st

*Communauté Financière Africaine franc

Africa continued		Area sq km	Area sq miles	Population	Capital	Languages	Religions	Currency	Internet Link
SENEGAL		196 720	75 954	14 133 000	Dakar	French, Wolof, Fulani, other local languages	Sunni Muslim, Roman Catholic, traditional beliefs	CFA franc*	www.gouv.sn
SEYCHELLES		455	176	93 000	Victoria	English, French, creole	Roman Catholic, Protestant	Seychelles rupee	www.virtualseychelles.sc
SIERRA LEONE		71 740	27 699	6 092 000	Freetown	English, creole, Mende, Temne, other local languages	Sunni Muslim, traditional beliefs	Leone	www.statehouse.gov.sl
SOMALIA		637 657	246 201	10 496 000	Mogadishu	Somali, Arabic	Sunni Muslim	Somali shilling	www.somaligov.net
SOUTH AFRICA		1 219 090	470 693	52 776 000	Pretoria/Cape Town/Bloemfontein	Afrikaans, English, nine official other local languages	Protestant, Roman Catholic, Sunni Muslim, Hindu	Rand	www.gov.za
SOUTH SUDAN		644 329	248 775	11 296 000	Juba	Arabic, Dinka, Nubian, Beja, English, other local languages	Christian, Sunni Muslim, traditional beliefs	South Sudanese pound	www.goss.org
SUDAN		1 861 484	718 725	37 964 000	Khartoum	Arabic, English, Nubian, Beja, Fur, other local languages	Sunni Muslim, traditional beliefs, Christian	Sudanese pound (Sudani)	www.presidency.gov.sd
TANZANIA		945 087	364 900	49 253 000	Dodoma	Swahili, English, Nyamwezi, other local languages	Shi'a Muslim, Sunni Muslim, traditional beliefs, Christian	Tanzanian shilling	www.tanzania.go.tz
TOGO		56 785	21 925	6 817 000	Lomé	French, Ewe, Kabre, other local languages	Traditional beliefs, Christian, Sunni Muslim	CFA franc*	www.republicoftogo.com
TUNISIA		164 150	63 379	10 997 000	Tunis	Arabic, French	Sunni Muslim	Tunisian dinar	www.ministeres.tn
UGANDA		241 038	93 065	37 579 000	Kampala	English, Swahili, Luganda, other local languages	Roman Catholic, Protestant, Sunni Muslim, traditional beliefs	Ugandan shilling	www.statehouse.go.ug
ZAMBIA		752 614	290 586	14 539 000	Lusaka	English, Bemba, Nyanja, Tonga, other local languages	Christian, traditional beliefs	Zambian kwacha	www.parliament.gov.zm
ZIMBABWE		390 759	150 873	14 150 000	Harare	16 official languages including English, Shona and Ndebele	Christian, traditional beliefs	US dollar and other currencies	www.zim.gov.zw

*Communauté Financière Africaine franc

Oceania		Area sq km	Area sq miles	Population	Capital	Languages	Religions	Currency	Internet Link
AUSTRALIA		7 692 024	2 969 907	23 343 000	Canberra	English, Italian, Greek	Protestant, Roman Catholic, Orthodox	Australian dollar	www.australia.gov.au
FIJI		18 330	7 077	881 000	Suva	English, Fijian, Hindi	Christian, Hindu, Sunni Muslim	Fijian dollar	www.fiji.gov.fj
KIRIBATI		717	277	102 000	Bairiki, Tarawa	Gilbertese, English	Roman Catholic, Protestant	Australian dollar	www.parliament.gov.ki
MARSHALL ISLANDS		181	70	53 000	Delap-Uliga-Djarrit	English, Marshallese	Protestant, Roman Catholic	United States dollar	www.rmi-op.net
MICRONESIA, FEDERATED STATES OF		701	271	104 000	Palikir	English, Chuukese, Pohnpeian, other local languages	Roman Catholic, Protestant	United States dollar	www.fsmgov.org
NAURU		21	8	10 000	Yaren (de facto capital)	Nauruan, English	Protestant, Roman Catholic	Australian dollar	www.naurugov.nr
NEW ZEALAND		270 534	104 454	4 506 000	Wellington	English, Maori	Protestant, Roman Catholic	New Zealand dollar	www.govt.nz
PAPUA NEW GUINEA		462 840	178 704	7 321 000	Port Moresby	English, Tok Pisin (creole), other local languages	Protestant, Roman Catholic, traditional beliefs	Kina	www.pm.gov.pg
SAMOA		2 831	1 093	190 000	Apia	Samoan, English	Protestant, Roman Catholic	Tala	www.samoagovt.ws
SOLOMON ISLANDS		28 370	10 954	561 000	Honiara	English, creole, other local languages	Protestant, Roman Catholic	Solomon Islands dollar	www.pmc.gov.sb
TONGA		748	289	105 000	Nuku'alofa	Tongan, English	Protestant, Roman Catholic	Pa'anga	www.tongaportal.gov.to
TUVALU		25	10	10 000	Vaiaku, Funafuti	Tuvaluan, English	Protestant	Australian dollar	
VANUATU		12 190	4 707	253 000	Port Vila	English, Bislama (creole), French	Protestant, Roman Catholic, traditional beliefs	Vatu	www.governmentofvanuatu.gov.vu

North America		Area sq km	Area sq miles	Population	Capital	Languages	Religions	Currency	Internet Link
ANTIGUA AND BARBUDA		442	171	90 000	St John's	English, creole	Protestant, Roman Catholic	East Caribbean dollar	www.ab.gov.ag
THE BAHAMAS		13 939	5 382	377 000	Nassau	English, creole	Protestant, Roman Catholic	Bahamian dollar	www.bahamas.gov.bs
BARBADOS		430	166	285 000	Bridgetown	English, creole	Protestant, Roman Catholic	Barbadian dollar	www.barbados.gov.bb
BELIZE		22 965	8 867	332 000	Belmopan	English, Spanish, Mayan, creole	Roman Catholic, Protestant	Belizean dollar	www.belize.gov.bz
CANADA		9 984 670	3 855 103	35 182 000	Ottawa	English, French, other local languages	Roman Catholic, Protestant, Eastern Orthodox, Jewish	Canadian dollar	www.canada.gc.ca
COSTA RICA		51 100	19 730	4 872 000	San José	Spanish	Roman Catholic, Protestant	Costa Rican colón	www.presidencia.go.cr
CUBA		110 860	42 803	11 266 000	Havana	Spanish	Roman Catholic, Protestant	Cuban peso	www.cubagob.gov.cu
DOMINICA		750	290	72 000	Roseau	English, creole	Roman Catholic, Protestant	East Caribbean dollar	www.dominica.gov.dm
DOMINICAN REPUBLIC		48 442	18 704	10 404 000	Santo Domingo	Spanish, creole	Roman Catholic, Protestant	Dominican peso	www.cig.gov.do
EL SALVADOR		21 041	8 124	6 340 000	San Salvador	Spanish	Roman Catholic, Protestant	United States dollar	www.presidencia.gob.sv
GRENADA		378	146	105 000	St George's	English, creole	Roman Catholic, Protestant	East Caribbean dollar	www.gov.gd
GUATEMALA		108 890	42 043	15 468 000	Guatemala City	Spanish, Mayan languages	Roman Catholic, Protestant	Quetzal	www.guatemala.gob.gt
HAITI		27 750	10 714	10 317 000	Port-au-Prince	French, creole	Roman Catholic, Protestant, Voodoo	Gourde	http://primature.gouv.ht
HONDURAS		112 088	43 277	8 098 000	Tegucigalpa	Spanish, Amerindian languages	Roman Catholic, Protestant	Lempira	http://congresonacional.hn/
JAMAICA		10 991	4 244	2 784 000	Kingston	English, creole	Protestant, Roman Catholic	Jamaican dollar	http://jis.gov.jm
MEXICO		1 972 545	761 604	122 332 000	Mexico City	Spanish, Amerindian languages	Roman Catholic, Protestant	Mexican peso	www.presidencia.gob.mx
NICARAGUA		130 000	50 193	6 080 000	Managua	Spanish, Amerindian languages	Roman Catholic, Protestant	Córdoba	www.presidencia.gob.ni
PANAMA		77 082	29 762	3 864 000	Panama City	Spanish, English, Amerindian languages	Roman Catholic, Protestant, Sunni Muslim	Balboa	www.presidencia.gob.pa
ST KITTS AND NEVIS		261	101	54 000	Basseterre	English, creole	Protestant, Roman Catholic	East Caribbean dollar	www.gov.kn
ST LUCIA		616	238	182 000	Castries	English, creole	Roman Catholic, Protestant	East Caribbean dollar	www.stlucia.gov.lc
ST VINCENT AND THE GRENADINES		389	150	109 000	Kingstown	English, creole	Protestant, Roman Catholic	East Caribbean dollar	www.gov.vc
TRINIDAD AND TOBAGO		5 130	1 981	1 341 000	Port of Spain	English, creole, Hindi	Roman Catholic, Hindu, Protestant, Sunni Muslim	Trinidad and Tobago dollar	www.ttconnect.gov.tt
UNITED STATES OF AMERICA		9 826 635	3 794 085	320 051 000	Washington D.C.	English, Spanish	Protestant, Roman Catholic, Sunni Muslim, Jewish	United States dollar	www.usa.gov

South America		Area sq km	Area sq miles	Population	Capital	Languages	Religions	Currency	Internet Link
ARGENTINA		2 766 889	1 068 302	41 446 000	Buenos Aires	Spanish, Italian, Amerindian languages	Roman Catholic, Protestant	Argentinian peso	www.argentina.gov.ar
BOLIVIA		1 098 581	424 164	10 671 000	La Paz/Sucre	Spanish, Quechua, Aymara	Roman Catholic, Protestant, Baha'i	Boliviano	www.bolivia.gob.bo
BRAZIL		8 514 879	3 287 613	200 362 000	Brasília	Portuguese	Roman Catholic, Protestant	Real	www.brazil.gov.br
CHILE		756 945	292 258	17 620 000	Santiago	Spanish, Amerindian languages	Roman Catholic, Protestant	Chilean peso	www.gob.cl
COLOMBIA		1 141 748	440 831	48 321 000	Bogotá	Spanish, Amerindian languages	Roman Catholic, Protestant	Colombian peso	www.gobiernoenlinea.gov.co
ECUADOR		272 045	105 037	15 738 000	Quito	Spanish, Quechua, other Amerindian languages	Roman Catholic	US dollar	www.presidencia.gob.ec
GUYANA		214 969	83 000	800 000	Georgetown	English, creole, Amerindian languages	Protestant, Hindu, Roman Catholic, Sunni Muslim	Guyanese dollar	www.gina.gov.gy
PARAGUAY		406 752	157 048	6 802 000	Asunción	Spanish, Guaraní	Roman Catholic, Protestant	Guaraní	www.presidencia.gov.py
PERU		1 285 216	496 225	30 376 000	Lima	Spanish, Quechua, Aymara	Roman Catholic, Protestant	Nuevo sol	www.peru.gob.pe
SURINAME		163 820	63 251	539 000	Paramaribo	Dutch, Surinamese, English, Hindi	Hindu, Roman Catholic, Protestant, Sunni Muslim	Surinamese dollar	www.president.gov.sr
URUGUAY		176 215	68 037	3 407 000	Montevideo	Spanish	Roman Catholic, Protestant, Jewish	Uruguayan peso	www.presidencia.gub.uy
VENEZUELA		912 050	352 144	30 405 000	Caracas	Spanish, Amerindian languages	Roman Catholic, Protestant	Bolívar	www.presidencia.gob.ve

National Statistics

The current pattern of the world's countries and territories is a result of a long history of exploration, colonialism, conflict and politics. The fact that there are currently 196 independent countries in the world – the most recent, South Sudan, only being created in July 2011 – illustrates the significant political changes which have occurred since 1950 when there were only eighty-two. There has been a steady progression away from colonial influences over the last fifty years, although many dependent overseas territories remain.

The shapes of countries and the pattern of international boundaries reflect both physical and political processes. Some borders follow natural features – rivers, mountain ranges, etc – others are defined according to political agreement or as a result of war. Some are still subject to dispute between two or more countries, and many remain undefined on the ground.

Facts

- The longest single continuous land border stretches for 6 416 kilometres between Canada and the USA

- Both China and Russia have land borders with 14 different countries

- Vatican City, the smallest independent country, was created in 1929 as an enclave within Rome, the capital of Italy

- All countries of the world are members of the United Nations except Kosovo, Taiwan and Vatican City

Internet Links

United Nations	www.un.org
Foreign and Commonwealth Office	www.fco.gov.uk
International Boundaries Research Unit	www.dur.ac.uk/ibru
Permanent Committee on Geographical Names	www.pcgn.org.uk
U.S. Board on Geographic Names	geonames.usgs.gov

Aerial view of the **Vatican City**, the world's smallest country by both population and area.

International boundaries in the sea shown on this map indicate ownership of islands and island groups only. They do not infer the alignments of legal maritime boundaries.

World extremes

Countries			
Largest country (area)	**Russia**	17 075 400 sq km	6 592 849 sq miles
Smallest country (area)	**Vatican City**	0.5 sq km	0.2 sq miles
Largest country (population)	**China**	1 369 993 000	
Smallest country (population)	**Vatican City**	800	
Most densely populated country	**Monaco**	19 000 per sq km	38 000 per sq mile
Least densely populated country	**Mongolia**	1.8 per sq km	4.7 per sq mile
Capitals			
Largest national capital (population)	**Tōkyō, Japan**	38 197 000	
Smallest national capital (population)	**Ngerulmud, Palau**	391	
Most northerly national capital	**Reykjavík, Iceland**	64° 08'N	
Most southerly national capital	**Wellington, New Zealand**	41° 18'S	
Highest national capital	**La Paz, Bolivia**	3 636 m	11 910 ft

World
Landscapes

The earth's physical features, both on land and on the sea bed, closely reflect its geological structure. The current shapes of the continents and oceans have evolved over millions of years. Movements of the tectonic plates which make up the earth's crust have created some of the best-known and most spectacular features. The processes which have shaped the earth continue today with earthquakes, volcanoes, erosion, climatic variations and man's activities all affecting the earth's landscapes.

The total topographic range of the earth's surface is nearly 20 000 metres, from the highest point Mount Everest, to the lowest point in the Mariana Trench. Major mountain ranges include the Himalaya, the Andes and the Rocky Mountains, each of which give rise to some of the world's greatest rivers. In contrast, the deserts of the Sahara, Australia, the Arabian Peninsula and the Gobi cover vast areas and each provide unique landscapes.

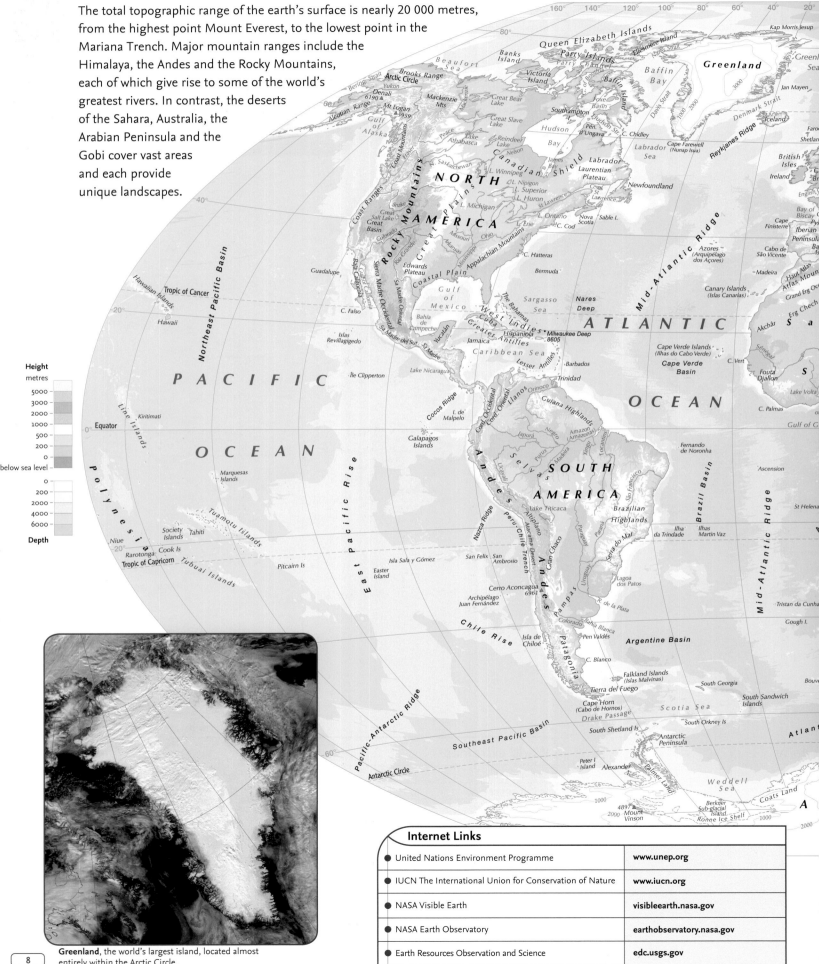

Height
metres
- 5000
- 3000
- 2000
- 1000
- 500
- 200
- 0
- below sea level
- 0
- 200
- 2000
- 4000
- 6000

Depth

Greenland, the world's largest island, located almost entirely within the Arctic Circle.

Internet Links	
● United Nations Environment Programme	**www.unep.org**
● IUCN The International Union for Conservation of Nature	**www.iucn.org**
● NASA Visible Earth	**visibleearth.nasa.gov**
● NASA Earth Observatory	**earthobservatory.nasa.gov**
● Earth Resources Observation and Science	**edc.usgs.gov**

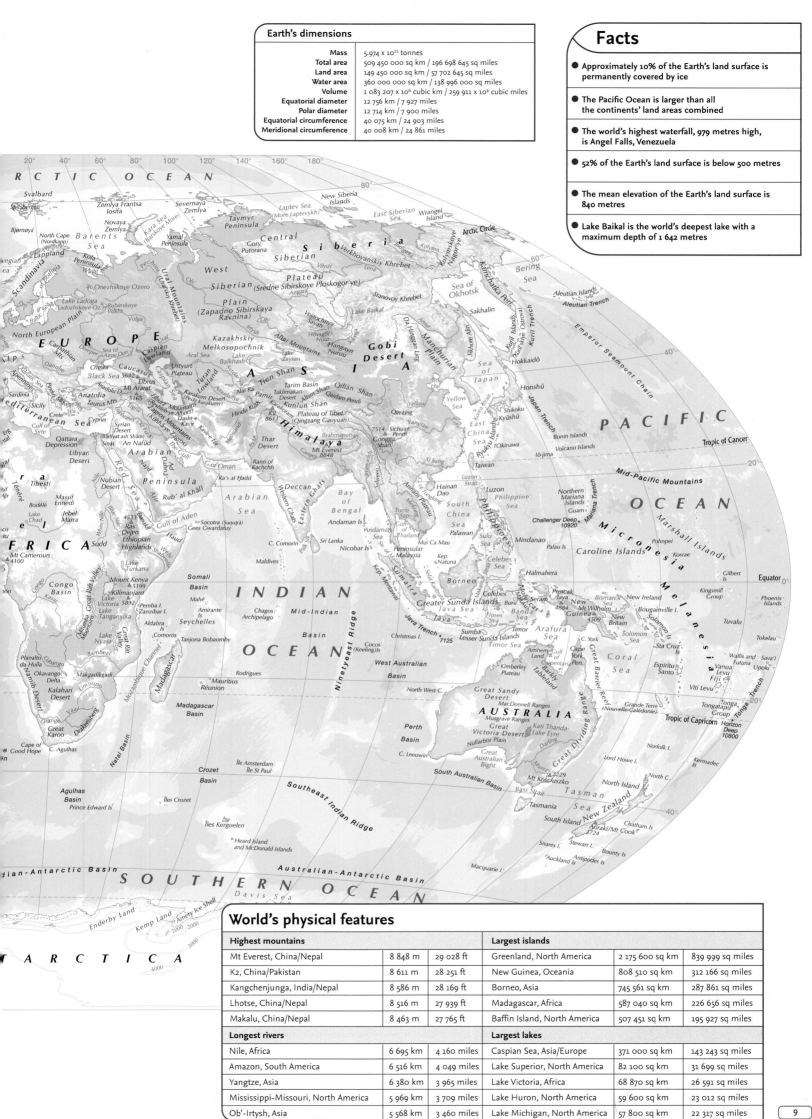

Earth's dimensions

Mass	5.974 x 10²¹ tonnes
Total area	509 450 000 sq km / 196 698 645 sq miles
Land area	149 450 000 sq km / 57 702 645 sq miles
Water area	360 000 000 sq km / 138 996 000 sq miles
Volume	1 083 207 x 10⁶ cubic km / 259 911 x 10⁶ cubic miles
Equatorial diameter	12 756 km / 7 927 miles
Polar diameter	12 714 km / 7 900 miles
Equatorial circumference	40 075 km / 24 903 miles
Meridional circumference	40 008 km / 24 861 miles

Facts

- Approximately 10% of the Earth's land surface is permanently covered by ice
- The Pacific Ocean is larger than all the continents' land areas combined
- The world's highest waterfall, 979 metres high, is Angel Falls, Venezuela
- 52% of the Earth's land surface is below 500 metres
- The mean elevation of the Earth's land surface is 840 metres
- Lake Baikal is the world's deepest lake with a maximum depth of 1 642 metres

World's physical features

Highest mountains			Largest islands		
Mt Everest, China/Nepal	8 848 m	29 028 ft	Greenland, North America	2 175 600 sq km	839 999 sq miles
K2, China/Pakistan	8 611 m	28 251 ft	New Guinea, Oceania	808 510 sq km	312 166 sq miles
Kangchenjunga, India/Nepal	8 586 m	28 169 ft	Borneo, Asia	745 561 sq km	287 861 sq miles
Lhotse, China/Nepal	8 516 m	27 939 ft	Madagascar, Africa	587 040 sq km	226 656 sq miles
Makalu, China/Nepal	8 463 m	27 765 ft	Baffin Island, North America	507 451 sq km	195 927 sq miles

Longest rivers			Largest lakes		
Nile, Africa	6 695 km	4 160 miles	Caspian Sea, Asia/Europe	371 000 sq km	143 243 sq miles
Amazon, South America	6 516 km	4 049 miles	Lake Superior, North America	82 100 sq km	31 699 sq miles
Yangtze, Asia	6 380 km	3 965 miles	Lake Victoria, Africa	68 870 sq km	26 591 sq miles
Mississippi-Missouri, North America	5 969 km	3 709 miles	Lake Huron, North America	59 600 sq km	23 012 sq miles
Ob'-Irtysh, Asia	5 568 km	3 460 miles	Lake Michigan, North America	57 800 sq km	22 317 sq miles

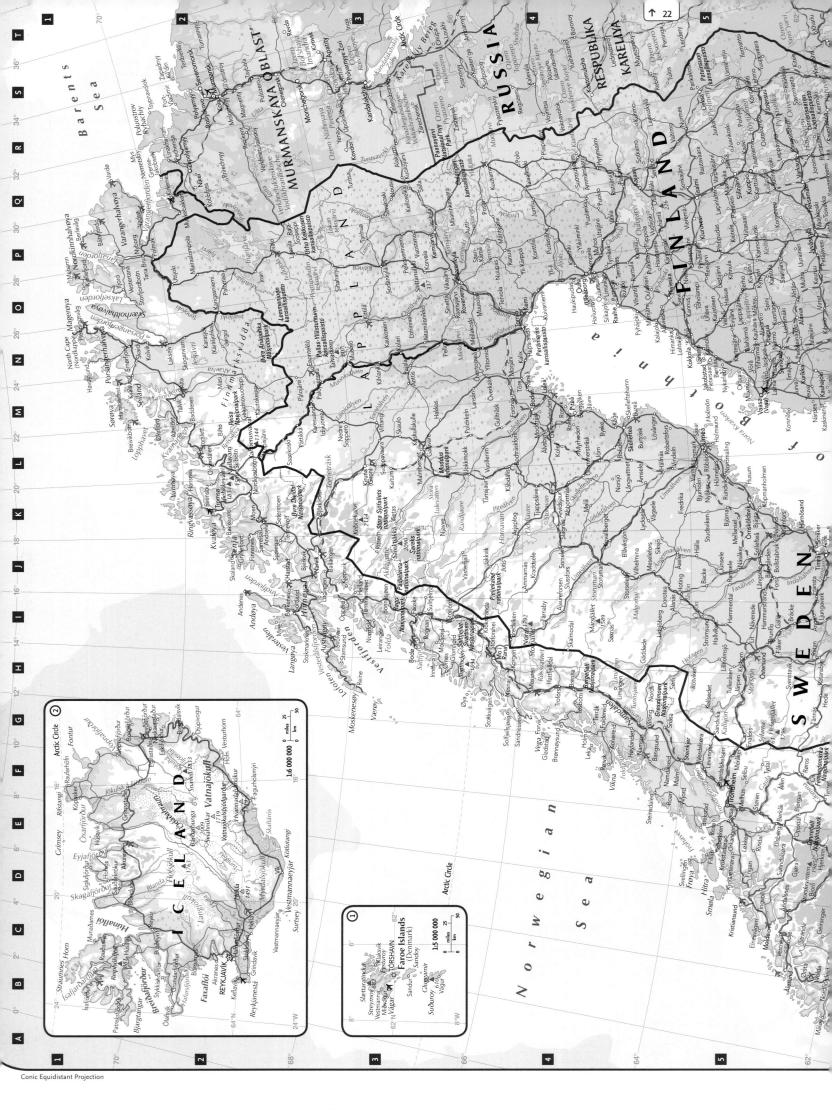

Conic Equidistant Projection

1:5 000 000

0 · 50 · 100 · 150 miles

0 · 50 · 100 · 150 · 200 · 250 km

Inset ② Iceland

Arctic Circle

1:6 000 000

0 · 25 · 50 miles

0 · 50 km

ICELAND

Vatnajökull

Inset ① Faroe Islands (Denmark)

Arctic Circle

1:5 000 000

0 · 25 · 50 miles

0 · 50 km

TÓRSHAVN

Europe
Scandinavia and the Baltic States

Europe
Northwest Europe

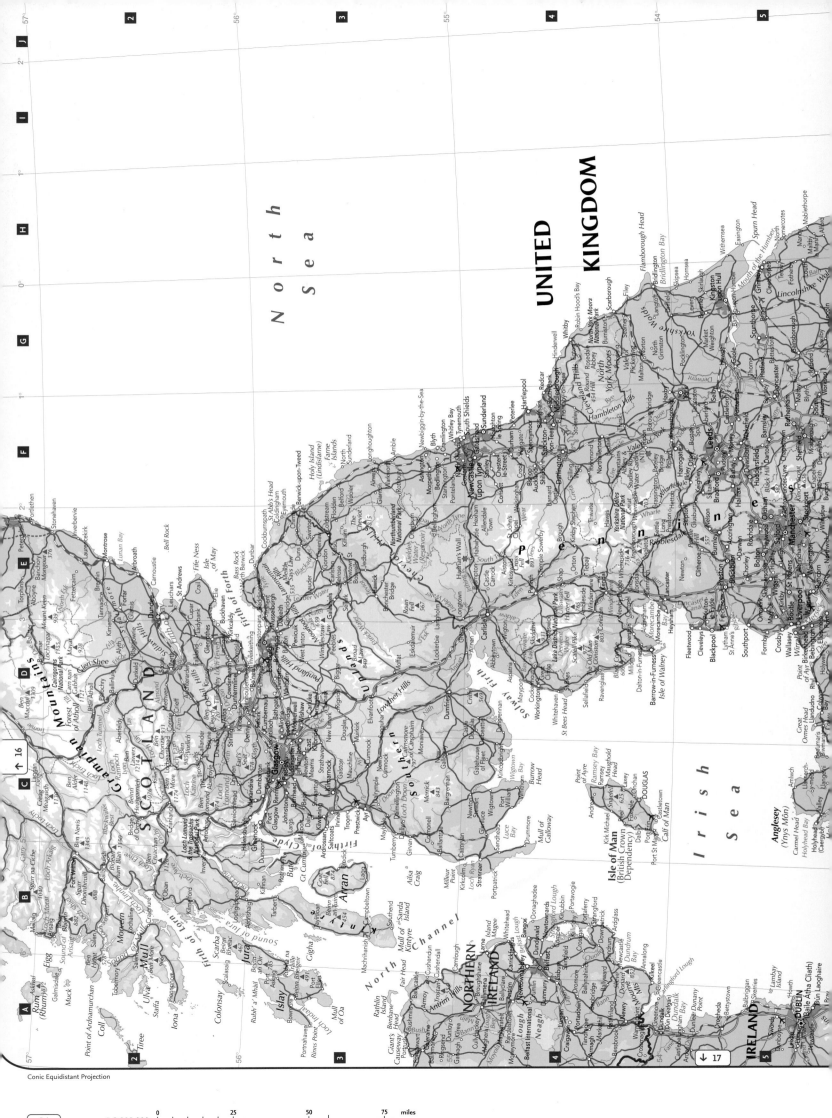

Conic Equidistant Projection

1:2 000 000

0   25   50   75 miles
0   25 50 75 100 125 km

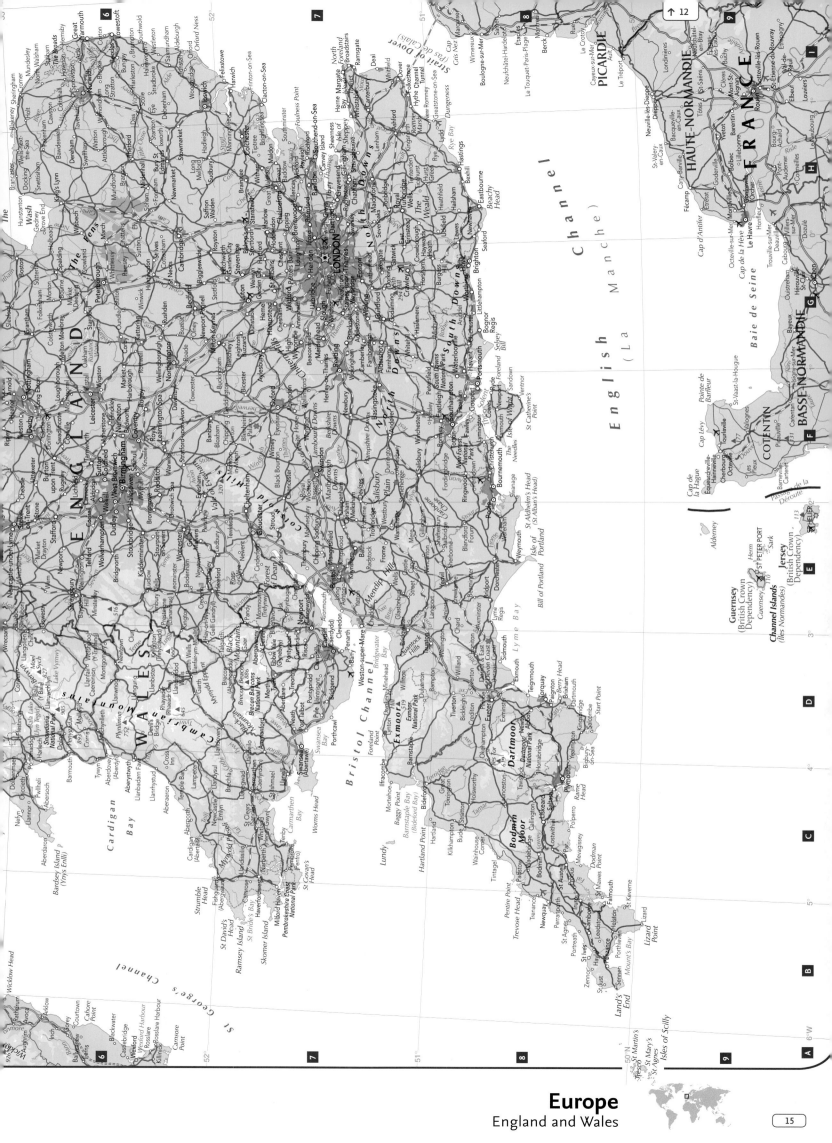

Europe

England and Wales

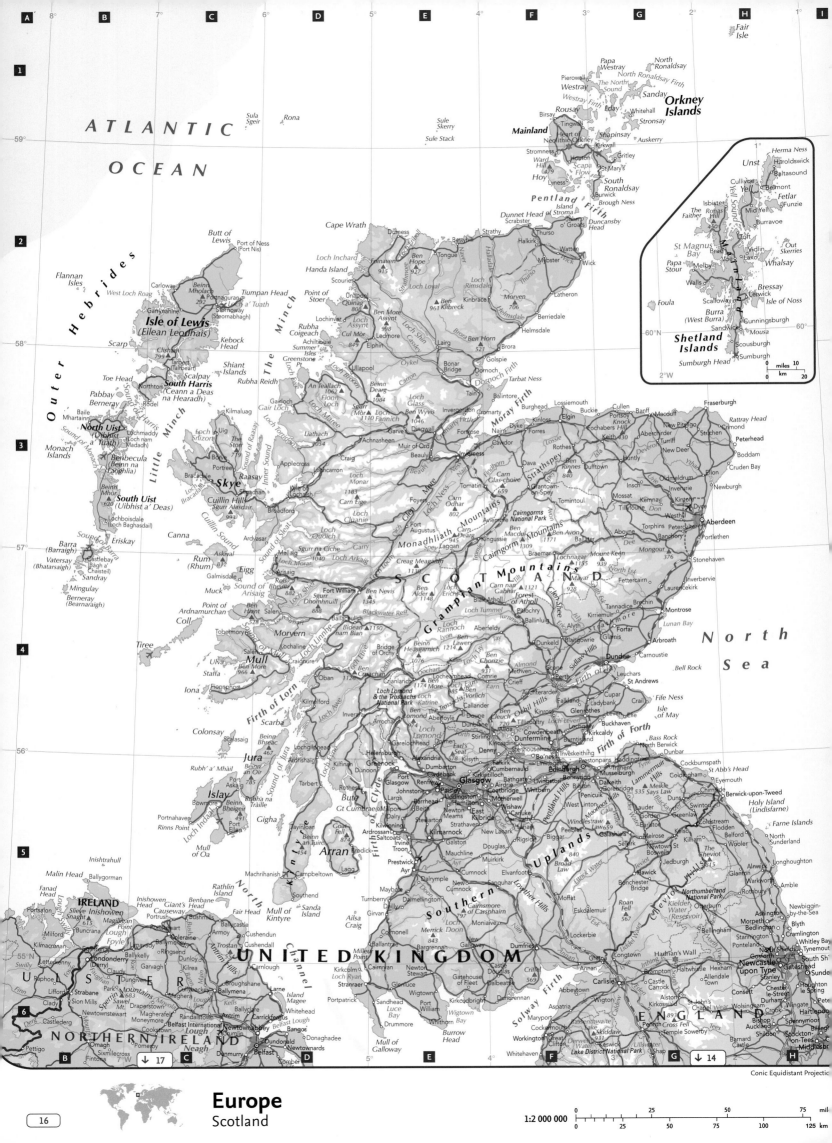

Europe
Scotland

16

1:2 000 000

Conic Equidistant Projection

Conic Equidistant Projection

1:2 000 000

| 0 | 25 | 50 | 75 | miles |
| 0 | 25 | 50 | 75 | 100 | 125 km |

Europe
Ireland

Conic Equidistant Projection

Europe
France

1:5 000 000

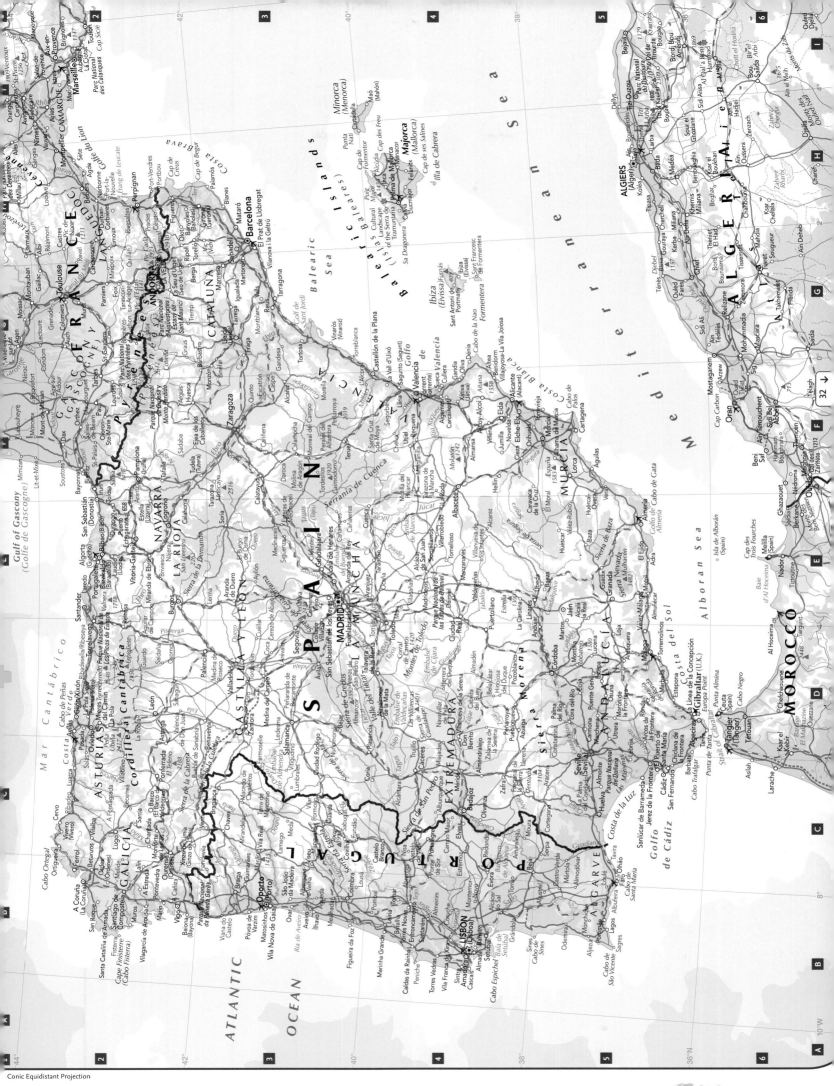

1:5 000 000

Conic Equidistant Projection

| 0 | 50 | 100 | 150 | miles |

| 0 | 50 | 100 | 150 | 200 | 250 | km |

→ 26

Europe
Italy and the Balkans

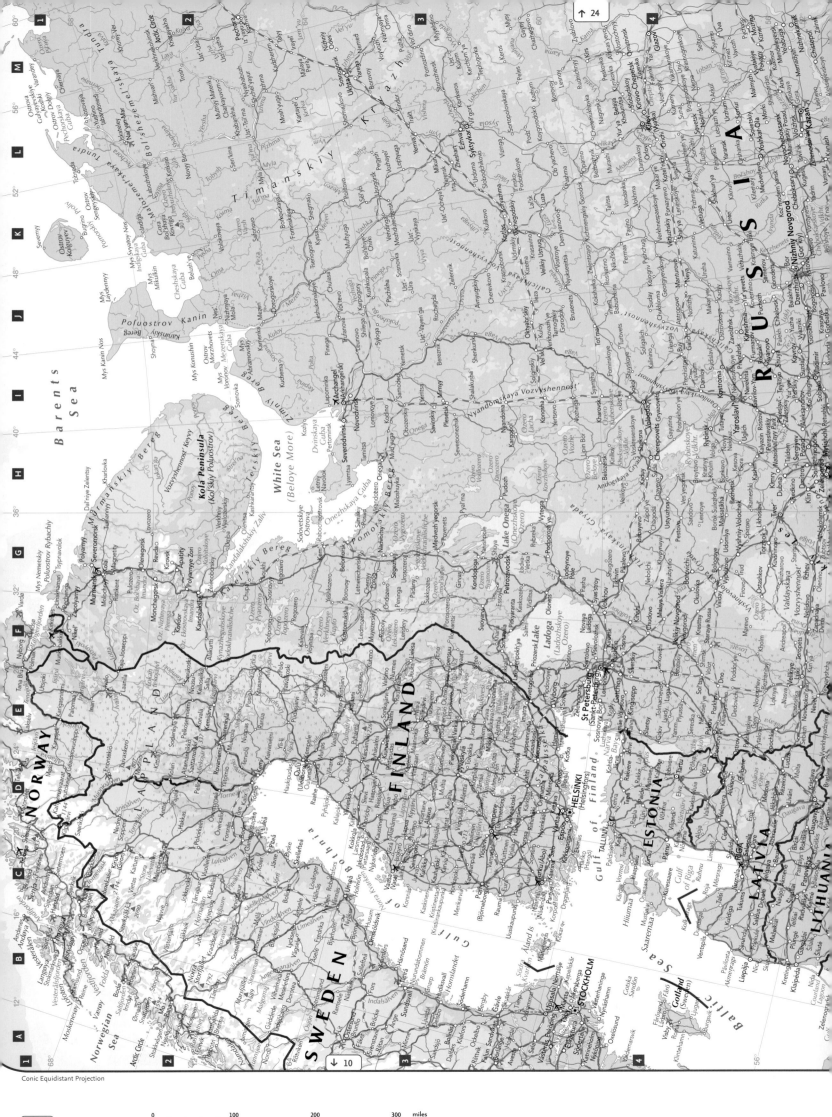

Conic Equidistant Projection

1:7 500 000

Europe
Western Russia

Conic Equidistant Projection

1:20 000 000

| 0 | 200 | 400 | 600 | miles |

| 0 | 200 | 400 | 600 | 800 | 1000 | km |

Asia
Northern Asia

Albers Conic Equal Area Projection

1:20 000 000

| 0 | 200 | 400 | 600 | miles |
| 0 | 200 | 400 | 600 | 800 | 1000 km |

Asia
Central and Southern Asia

Albers Conic Equal Area Projection

1:20 000 000

Asia
Eastern and Southeast Asia

Conic Equidistant Projection

1:7 000 000

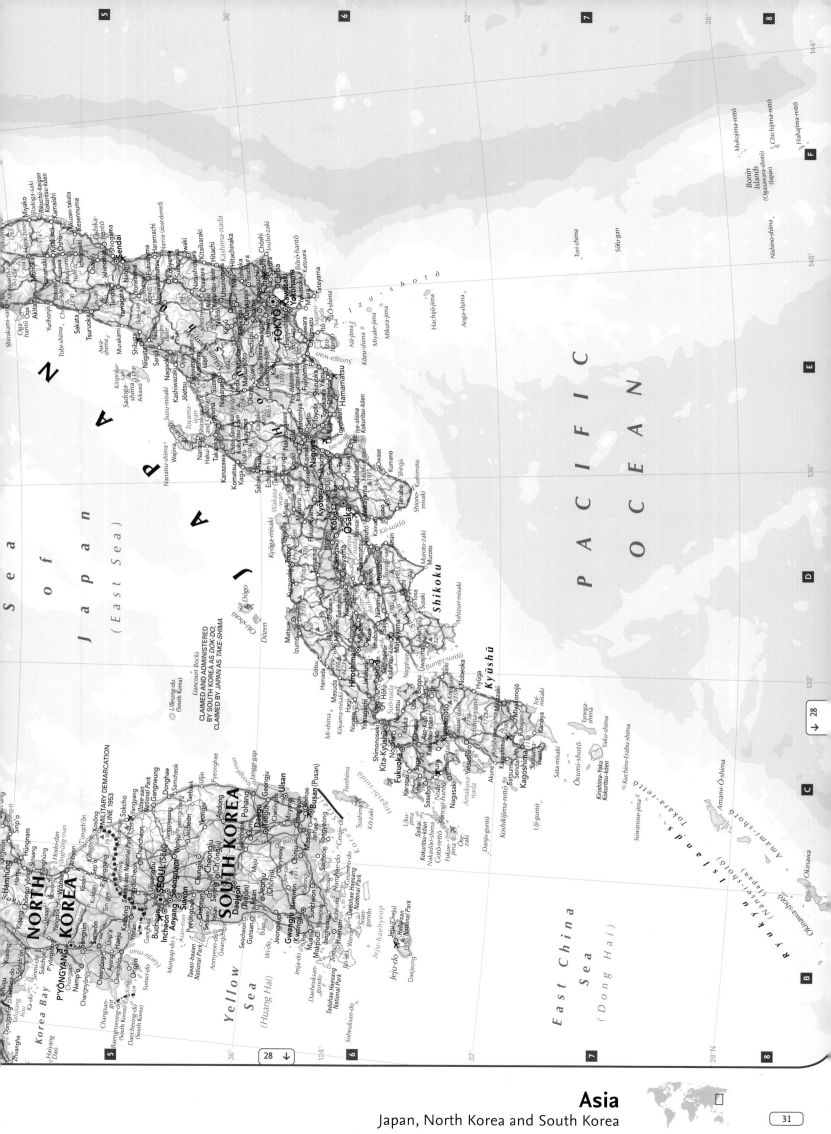

Asia
Japan, North Korea and South Korea

ATLANTIC OCEAN

MOROCCO

WESTERN SAHARA
ADMINISTERED BY MOROCCO

Canary Islands (Spain)

Madeira (Portugal)

Arquipélago da Madeira

Ilha de Porto Santo

FUNCHAL

La Palma
La Gomera
El Hierro
Santa Cruz de la Palma
Santa Cruz de Tenerife
Gran Canaria
Las Palmas de Gran Canaria
Lanzarote
Fuerteventura

SPAIN
Gibraltar

RABAT
Casablanca

ALGERIA

Grand Erg Occidental
Grand Erg Oriental
Plateau du Tademaït
Hamada de Tinrhert

TUNISIA
ALGIERS (Alger)

TRIPOLITA

Tropic of Cancer

S A H A R A

Erg Chech

MAURITANIA
EL MREYYÉ

NOUAKCHOTT

AZAWAD

MALI

NIGER

SENEGAL
DAKAR

THE GAMBIA
BANJUL

GUINEA BISSAU
BISSAU

Arquipélago dos Bijagós

GUINEA
CONAKRY

Timbuktu (Tombouctou)
Gao

NIAMEY

BURKINA FASO
OUAGADOUGOU

SIERRA LEONE
FREETOWN

Sherbro Island

CÔTE D'IVOIRE (IVORY COAST)

BENIN

GHANA
ACCRA

TOGO
LOMÉ

NIGERIA
ABUJA
Lagos
PORTO-NOVO

LIBERIA
MONROVIA

YAMOUSSOUKRO
Abidjan

Gulf of Guinea

Bight of Benin

Slave Coast

CAMEROO
YAOUNDÉ
Douala
MALABO

EQUATORIAL GUINEA
Bioko

SÃO TOMÉ AND PRÍNCIPE
SÃO TOMÉ
Príncipe

GABON
LIBREVILLE

Annobón (Equatorial Guinea)

ATLANTIC OCEAN

CAPE VERDE (CABO VERDE)

Ilhas do Cabo Verde

Santo Antão
Mindelo
São Vicente
São Nicolau
Porto Novo
Sal
Santa Maria
Boa Vista

Santiago (São Tiago)
Tarrafal
Maio
Brava
Fogo
2829
PRAIA

Equator

1:16 000 000

miles 100
km 150

Lambert Azimuthal Equal Area Projection

miles
0   200   400
km
0   200   400   600   800

Africa
Northern Africa

Lambert Azimuthal Equal Area Projection

1:16 000 000

Africa
Central and Southern Africa

GHANZI
BOTSWANA
Central Kalahari Game Reserve
KGALAGADI
K a l a h a r i
Desert
KWENEN
SOUTHER

ERONGO
KHOMAS
OMAHEKE
WINDHOEK

NAMIBIA
HARDAP

Namib-
Naukluft
Park

Namib-Skeleton Coast National Park

GREAT NAMAQUALAND

!KARAS

Gemsbok National Park
Kalahari Gemsbok National Park
Kgalagadi Transfrontier Park

NORT

Sperrgebiet National Park

Ai-Ais Hot Springs and Fish River Canyon Park
Ai-Ais/Richtersveld Transfrontier Park

Richtersveld Cultural and Botanical Landscape

SOUT

NAMAQUALAND

NORTHERN
CAPE

AFRI

Goegap Nature Reserve
Namaqua National Park

ATLANTIC

OCEAN

Great Karoo

WESTERN CAPE

Little Karoo

CAPE TOWN

Table Mountain National Park
Cape of Good Hope

Tropic of Capricorn

Lambert Azimuthal Equal Area Projection

0   50   100   150   miles
0   50   100   150   200   250   km

B 120° **C** 140° **E** 150° **F**

Celebes Sea
Tanjungselor
Tanjungredeb
Morotai
Tobelo
Manado
Ternate Halmahera
Morotai

1

Sambaliung
Tolitoli
Gorontalo
Waigeo
Tanjung d'Urville
Sarmi
Wuvulu Island
Hermit Islands
St Matthias Group
Mussau Island
New Hanover
Lyra Reef

Borneo
Sangkulirang
Equator
Samarinda
Palu
Teluk Tomini
Kwoka
3000
Jazirah Doberai
Manokwari
Biak
Numfoor
Biak
Teluk Cenderawasih
Van Rees
Jayapura
Vanimo
Wewak
Schouten Islands
Manam Island
Karkar Island
Admiralty Islands
Manus Island
New Ireland
Tabar Islands
Lihir Group
Tanga Islands
Rabaul
Feni Islands
Sohano

Balikpapan
Poso
Luwuk
Banggai
Salawati Sorong
Misoöl
Teminabuan
Pegunungan Tariku
Taritatu
Sepik
Madang
New Britain
Bougainville Island
Treasury Is

Celebes (Sulawesi)
Kolonedale
Kepulauan Banggai
Wahai
Seram
3019
Laut Seram (Ceram Sea)
Fakfak
Kaimana
Nabire
Pegunungan Maoke
Tembagapura Puncak Jaya
4884
Mount Hagen
Goroka
Lae
NEW GUINEA
Long Island
Witu Islands
Umboi
Kimbe
New Britain
Solomon Sea

Mamuju
Makale
Mangole
Obi
Teluk Berau
Adi
Kamrau
Puncak Trikora
4730
Puncak Mandala
Mendi
Kikori
Kerema
Mount Victoria
Kokoda
Goodenough Island
D'Entrecasteaux Islands

2

Majene
Parepare
Kendari
Wowoni
Buru
Ambon Ambon
Kepulauan Watubela
Kai Besar
Dobo
Benjina
Wokam
Kepulauan Aru
Kobroör
Balimo
Gulf of Papua
Bereina
4077
Owen Stanley Range
Lae
Lusancay Islands and Reefs
Trobriand Islands
Woodlark Island

Makassar (Ujung Pandang)
Sinjai
Baubau
Buton
Kepulauan Tukangbesi
Laut Banda (Banda Sea)
Trangan
Larat
Kai Kecil
Tual
Sia
Tanjung Deyong
Pulau Dolok
Merauke
PORT MORESBY
Kwikila
Normanby Island
Conflict Group

Bontosunggu
Bulukumba
Benteng
Kepulauan Bonerate
Kepulauan Barat Daya
Damar
Kepulauan Tanimbar
Yamdena
Tanjung Vals
Torres Strait
Cape York
D'Entrecasteaux Islands
Tagula Island
Louisiade Archipelago
Rossel Island

Kepulauan Kangean
Pulau Selayar
Kepulauan Alor
Wetar
Kepulauan Leti
Selaru

INDONESIA

Kepulauan Tengah
Laut Flores (Flores Sea)
Kalabahi
Alor
Romang
Babar

Bali
Lombok
Dompu Raba
Ruteng
Flores
DILI
9450
Fobo
Fatamailau
EAST TIMOR
Kefamenanu
Timor
EAST TIMOR (TIMOR-LESTE)
Prince of Wales Island
Cape York

Denpasar
Mataram
Sumbawa
Ende
Laut Sawu (Savu Sea)
Kupang
Cape Wessel
Wessel Islands
Cape Grenville
Weipa
Cape York
Peninsula
Coen
Princess Charlotte Bay
Cape Melville
Osprey Reef

INDIAN OCEAN
Waikabubak Waingapu
Rote
Timor Sea
Melville Island
Cobourg Peninsula
Croker Island
Goulburn Islands
Cape Arnhem
Gulf of Carpentaria
Albatross Bay
Laura
Cape Flattery

Sumba
Bathurst Island
Van Diemen Gulf
Jabiru
Arnhem Land
Alyangula
Groote Eylandt
Mitchell
Cooktown
Mossman
Cairns

Ashmore and Cartier Islands (Australia)
Cape Londonderry
Joseph Bonaparte Gulf
Beagle Gulf
Darwin
Rum Jungle (abandoned)
Adelaide River
Pine Creek
Matanaka
Sir Edward Pellew Group
Mornington Island
Wellesley Islands
Normanton
Innisfail
Tully

3

Bonaparte Archipelago
Cape Lévêque
Collier Bay
King Sound
Derby
Cape Leveque
Kimberley Plateau
Mount Ord 936
Halls Creek
Timber Creek
Daly Waters
Victoria River Downs
Larrimah
Borroloola
Barkly Tableland
Forsayth
Townsville
Ayr
Whitsunday Group
Bowen

Broome
Liveringa
Fitzroy Crossing
King Leopold Ra.
Lajamanu
Lake Argyle
Wyndham
Kununurra
Dryack
Daly
NORTHERN
Lake Woods
Camooweal
Kajabbi
Gregory Range
Charters Towers
1277
Mackay
Sarina
Swain Reefs

Roebuck Bay
Eighty Mile Beach
Great Sandy Desert
Lake White
Tennant Creek
TERRITORY
Barrow Creek
Mount Isa
Cloncurry
Flinders
Winton
Hughenden
Clermont
Yeppoon
Curtis I.
Gladstone

20°
Port Hedland
Shay Gap (abandoned)
Nullagine
Lake Mackay
Lake Wills
Yuendumu
Tanami Desert
Alice Springs
Dajarra
Boulia
QUEENSLAND
Longreach
Blackall
Bardaline
Emerald
Rockhampton
Sandy C.
Bundaberg
Maryborough

Barrow Island
Karratha Roebourne
Marble Bar
PILBARA
Chichester Range
Newman
Lake Macdonald
Mount Liebig 1524
Mount Zeil 1531
MacDonnell Ranges
Erldunda
Simpson Desert
Birdsville
Bilpa Morea Claypan
Georgina
Yaraka
Windorah
Charleville
Mitchell
Roma
Kingaroy
Gympie
Nambour
Caboolture

North West Cape
Exmouth Gulf
Onslow
Pannawonica
Hamersley Range
1249
Mount Meharry
Paraburdoo
Newman
Lake Disappointment
Gibson Desert
Lake Hopkins
Lake Neale
Lake Amadeus
Uluru/Ayers Rock 863
Everard Range
Oodnadatta
Sturt Stony Desert
Coopers Creek
Cunnamulla
Quilpie
Dirranbandi
Darling Downs
Toowoomba
Brisbane
Beenleigh
Gold Coast

Coral Bay
Tropic of Capricorn
Minilya
Mount Augustus 1106
Ashburton
Robinson Ranges
Wiluna
Great Victoria Desert
Warburton
Musgrave Ranges
Mount Woodroffe 1440
Marla
Kati Thanda–Lake Eyre (North)
Strzelecki Desert
Hungerford
Bourke
Brewarrina
Moree
Inverell
Grafton
Byron Bay
Ballina

4
Shark Bay
Lake MacLeod
Denham
Dirk Hartog Island
Carnarvon
Murchison
Meekatharra
Robinson Ranges
Lake Carnegie
Lake Wells
AUSTRALIA
WESTERN
Lake Maurice
SOUTH
AUSTRALIA
Lake Blanche
Tibooburra
Wilcannia
Cobar
Warren
Narrabri
Armidale
Macksville
Coffs Harbour
Port Macquarie

Kalbarri
Northampton
Mullewa
Mount Magnet
Menzies
Leonora
Laverton
Lake Carey
Maralinga
Lake Torrens
Woomera
Flinders Ranges
Lake Frome
Broken Hill
Lake Menindee
Ivanhoe
Dubbo
Orange
Tamworth
Taree

Houtman Abrolhos
Geraldton
Dongara
Lake Moore
Lake Barlee
Lake Ballard
Kalgoorlie
Nullarbor Plain
Penong
Lake Gairdner
Port Augusta
Jamestown
Burra
Lake Yamma Yamma
Parkes
Forbes
Lithgow
Newcastle

110°
Mullewa
Moora
Bonnie Rock
Southern Cross
Coolgardie
Kambalda
Lake Cowan
Eucla
Mundrabilla
Fowlers Bay
Ceduna
Streaky Bay
Whyalla
Kyancutta
Port Pirie
Port Augusta
NEW SOUTH WALES
Griffith
Hay
Swan Hill
Orange
Sydney
Wollongong

30°
Mukinbudin
Merredin
Norseman
Balladonia
Great Australian Bight
Anxious Bay
Eyre Peninsula
Port Lincoln
Cowell
Yorke Pen.
Spencer Gulf
Adelaide
Murray Bridge
Mildura
Wagga Wagga
A.C.T.
CANBERRA
Narooma
JERVIS BAY TERR.

Perth
Fremantle
Rockingham
Mandurah
Bunbury
Busselton
Darling Range
Yanchep
Hyden
Esperance
Archipelago of the Recherche
Kangaroo Island
Cape Jaffa
Kingscote
Investigator Strait
Kangaroo Island
Goulburn
Bega
Eden
Mount Kosciuszko 2229
Great Dividing Range
Cape Howe

Margaret River
Geographe Bay
Cape Leeuwin
Katanning
Hood Point
Denmark
Albany
Cape Otway
Mount Gambier
VICTORIA
Ballarat
Bendigo
Shepparton
Wangaratta
Mount William 1167
Geelong
Melbourne
Sale
Bairnsdale
Ta

Point D'Entrecasteaux
Discovery Bay
Portland
Warrnambool
Horsham
Nhill
Lake Tyrrell
Bass Strait
Wilson's Promontory
Furneaux Group
Flinders Island

5
King Island
Currie
Hunter Islands
Burnie
Devonport
Eddystone Point
Launceston
Great Lake
Queenstown
TASMANIA
Lake Gordon
Hobart
Port Arthur

6
South East Cape

40°S

A 110°E **B** 120° **C** 130° **D** 140° **E** 150°

Oceania
Australia, New Zealand and Southwest Pacific

1:13 000 000

0   200   400   miles

0   200   400   600   800   km

Oceania
Australia

Oceania
Southeast Australia

Lambert Azimuthal Equal Area Projection

1:5 000 000

NEW ZEALAND

Tasman Sea

PACIFIC OCEAN

North Island
(Te Ika-a-Māui)

South Island
(Te Waipounamu)

Three Kings Islands
Cape Maria van Diemen
North Cape
Te Paki
Ninety Mile Beach
Cape Karikari
Doubtless Bay
Awanui
Ahipara Bay
Tauroa Point
Kaitaia
Kerikeri
Cape Brett
Broadwood
Russell
Taheke
Kawakawa
Poor Knights Islands
Donnellys Crossing
Whangarei
Dargaville
Mokohinau Islands
Maungaturoto
Bream Bay
Little Barrier Island
Port Fitzroy
Great Barrier Island
Wellsford
Leigh
Kawau Island
Colville Channel
North Head
Kaipara Harbour
Onewa
Haurakī Gulf
Mercury Islands
East Coast Bays
Whitianga
Takapuna
Waiheke Island
Colville
Coromandel Peninsula
Auckland
Manukau
The Aldermen Islands
Papakura
Whangamata
Manukau Harbour
Pukekohe
Waiuku
Thames
Mayor Island
Port Waikato
Huntly
Matakana Island
Cape Runaway
Tauranga
Whakatāne
Hicks Bay
Hamilton
Te Awamutu
Rotorua
Kawerau
Ōpōtiki
East Cape
Cambridge
Ruatoria
Kawhia Harbour
Kihikihi
Lake Rotorua
Te Teko
Raukumara Range
Te Kuiti
Ōtorohanga
Mount Tarawera
Te Urewera National Park
Tokomaru Bay
Piopio
Mangakino
Taupo
Matawai
Tolaga Bay
Awakino
Waitahanui
Gisborne
Mokau
Ōkahukura
Poverty Bay
New Plymouth
Ōhura
Tūrangi
Kaimanawa Mountains
Tarawera
Kaitawa
Wairoa
Mount Taranaki (Mount Egmont)
Whanganui National Park
Tongariro National Park
Mahaka
Table Cape
Cape Egmont
Egmont National Park
Ōhakune
Tokaanu
Taihape
Mahia Peninsula
Opunake
Raetihi
Ruahine Range
Hawke Bay
Hawera
Waiouru
Bay View
Napier
Patea
Taihape
Hastings
Cape Kidnappers
South Taranaki Bight
Wanganui
Havelock North
Waimarama
Turakina
Marton
Apiti
Waipawa
Feilding
Waipukurau
Palmerston North
Dannevirke
Porangahau
Levin
Cape Turnagain
Eketahuna
Pongaroa
Paraparaumu
Tararua Range
Upper Hutt
Masterton
Castlepoint
Kapiti Island
Porirua
Lower Hutt
WELLINGTON
Featherston
Lake Wairarapa
Mount Ross 981
Palliser Bay
Cape Palliser

Cape Farewell
Farewell Spit
Collingwood
Golden Bay
Cape Stephens
Kahurangi Point
Abel Tasman National Park
D'Urville Island
French Pass
Cook Strait
Tasman Mountains
Tasman Bay
Kahurangi National Park
Karamea
Riwaka
Kaikoura
Richmond
Nelson
Havelock
Picton
Waimangaroa
Seddonville
Owen River
Buller
Wakefield
Blenheim
Westport
Hope Saddle
Wairau
Charleston
Owen River
Seddon
Paparoa National Park
Reefton
Mount Travers 2338
Cape Campbell
Ahaura
Victoria Range
2121
Greymouth
Springs Junction
Nelson Lakes Nat. Park
Inland Kaikoura Range
Tapuae-o-Uenuku 2885
Lewis Pass
Hokitika
Rotomanu
Manakau 2608
Ross
Lake Sumner
Clarence
Ōtira
Lake Brunner
Arthur's Pass National Park
Hope
Kaikoura
Culverden
Parnassus
Abut Head
Harihari
Cheviot
Westland Tai Poutini National Park
Hari Hari
Waikari
Waipara
Pegasus Bay
Franz Josef Glacier
Mount Arrowsmith 2781
Hanmer
Fox Glacier 3117
Oxford
Kaiapoi
Aoraki/Mount Cook 3724
Rangiora
Lake Paringa
Aoraki/Mount Cook National Park
Sheffield
Christchurch
Jackson Head
Te Pirita
Sumner
Cascade Point
Mount Ward
Aylesbury
Banks Peninsula
Awarua Point
Mount Aspiring National Park
Lake Tekapo
Mayfield
Akaroa
Mount Aspiring 3033
L. Tekapo
Ashburton
Milford Sound
Mount Allan 2339
Fairlie
Longbeach
Milford Sound
Burkes Pass
Canterbury Bight
George Sound
Mount Cosmo 2476
Richardson Mts
Lake Hawea
Temuka
Secretary Island
Lake Pukaki
Geraldine
Doubtful Sound
Lake Wanaka
Wanaka
Twizel
The Hunters Hills
Timaru
Fiordland
Glenorchy
Mount Burns 1645
Pareora
Lake Te Anau
James Peak 2172
Hawkdun Range
Pukaki
Breaksea Sound
Dunstan Mts
Kurow
Waimate
Te Anau
Eyre Mountains
Ōtematata
Fiordland National Park
Mossburn
Ranfurly
Studholme Junction
Moeraki Point
Lake Manapouri
Ahol
Alexandra
Shag Point
Resolution Island
Roxbury
Hyde
Waikouaiti
Palmerston
Lumsden
Middlemarch
Warrington
Cape Providence
Caroline Peak 1704
Ohai
Mandeville
Beaumont
Mosgiel
Port Chalmers
Otago Peninsula
Puysegur Point
Winton
Gore
Waipahi
Otago Peninsula
Te Waewae Bay
Orepuki
Mataura
Dunedin
Edendale
Mount Pye 720
Waiwera
Henley
Riverton
Wyndham
Milton
Invercargill
Bluff
Kaitangata
Nugget Point
Fortrose
Long Beach
Foveaux Strait
Ruapuke Island
Chaslands Mistake
Solander Island
Codfish Island
Halfmoon Bay
Mason Bay
Rakiura National Park
Shelter Point
Titi Islands (Muttonbird Islands)
Stewart Island
South West Cape
North Trap

Conic Equidistant Projection

1:5 250 000

0   50   100   150 miles
0   50 100 150 200 250 km

Oceania
New Zealand

43

Lambert Conformal Conic Projection

1:16 000 000

| 0 | 200 | 400 | miles |
| 0 | 200 | 400 | 600 | 800 km |

North America
Canada

Lambert Conformal Conic Projection

1:12 000 000

North America
Northeast United States

Lambert Conformal Conic Projecti

1:3 500 000

North America
Southwest United States

Lambert Conformal Conic Projection

1:3 500 000

| 0 | 50 | 100 | miles |

| 0 | 50 | 100 | 150 | 200 | km |

Lambert Conformal Conic Projection

1:14 000 000

H WEST 75° I Richmond J 70° K 65° L 60° M 55° 1

35°

ATLANTIC

O C E A N 30°

HAMILTON Bermuda
(U.K.)

Tropic of Cancer

THE
BAHAMAS

W
e
s
t

I
n
d
i
e
s

HAVANA
(La Habana)

CUBA

Turks and
Caicos Islands
(U.K.)

GRAND TURK
(Cockburn Town)

Virgin
Islands
(U.K.) Anguilla
(U.K.) Leeward Islands

SAN
JUAN ROAD
TOWN St-Martin (France)
St-Barthelemy
(France) Barbuda
ANTIGUA
AND BARBUDA

CHARLOTTE
AMALIE Saba (Neth.)
St Eustatius (Neth.) BASSETERRE
ST JOHN'S Antigua

HAITI Hispaniola Virgin
Islands
(U.S.A.) St Kitts
AND NEVIS
Montserrat
(U.K.) Guadeloupe
(France)
BASSE-
TERRE

PORT-AU-
PRINCE SANTO
DOMINGO Puerto Rico
(U.S.A.) Les DOMINICA
ROSEAU

DOMINICAN
REPUBLIC Antilles FORT-DE-FRANCE
Martinique
(France)
CASTRIES ST LUCIA
BARBADOS

JAMAICA Lesser St Vincent Passage BRIDGETOWN

KINGSTON ST VINCENT
AND
THE GRENADINES
KINGSTOWN Windward Islands

C a r i b b e a n S e a GRENADA
ST GEORGE'S TRINIDAD
AND
TOBAGO
PORT
OF SPAIN

Lesser Antilles
Aruba
(Neth.) Curaçao
(Neth.) WILLEMSTAD
ORANJESTAD Bonaire
(Neth.)

VENEZUELA GUYANA

CARACAS

COLOMBIA

BOGOTÁ

BRAZIL

I 80° J 75° K 70° L

North America
Central America and the Caribbean

51

PACIFIC

OCEAN

Galapagos Islands
(Islas Galápagos)
(Ecuador)

Parque Nacional
Galápagos

Equator

Isla Santiago

Isla Fernandina

Isla Isabela

Isla Santa Cruz

Puerto Baquerizo Moreno

Isla San Cristóbal

Isla Floreana

1:14 000 000

miles 100

km 150

NICARAGUA
MANAGUA

COSTA RICA
SAN JOSÉ

PANAMA
PANAMA CITY

COLOMBIA
BOGOTÁ

VENEZUELA
CARACAS

GRENADA
ST GEORGE'S

TRINIDAD
AND
TOBAGO
PORT OF
SPAIN

ECUADOR
QUITO

PERU
LIMA

BOLIVIA
LA PAZ
SUCRE

CHILE

ARGENTINA

Lambert Azimuthal Equal Area Projection

1:14 000 000

miles 200 400

km 200 400 600 800

South America
Northern South America

South America
Southern South America

1:14 000 000

Lambert Azimuthal Equal Area Projection

Lambert Azimuthal Equal Area Projection

1:7 000 000

| 0 | 100 | 200 | miles |

| 0 | 100 | 200 | 300 | 400 km |

South America
Southeast Brazil

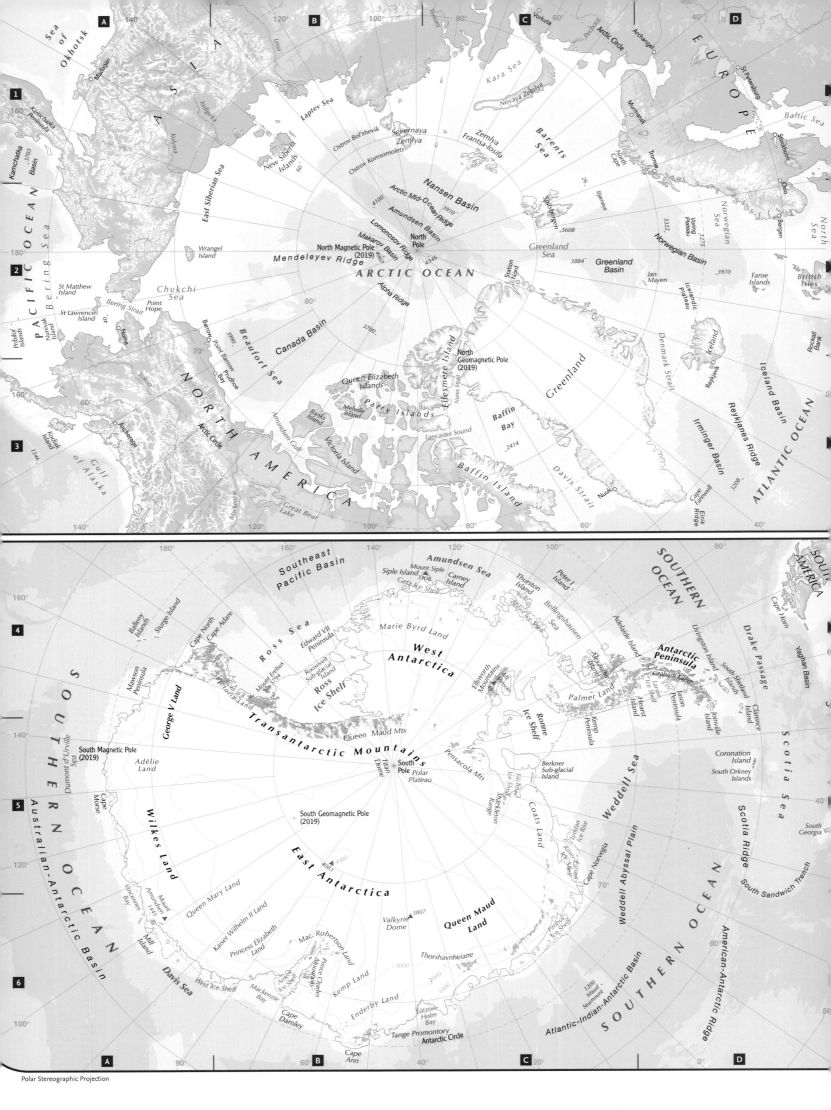

Arctic Ocean and Antarctica

Polar Stereographic Projection

1:35 000 000

Index

The index includes the most significant names on the maps in the atlas. The names are generally indexed to the largest scale map on which they appear. For large physical features this will be the largest scale map on which they appear in their entirety or in the majority. Names can be located using the grid reference letters and numbers around the edges of the map. Names located on insets have a symbol □.

Abbreviations used to describe features in the index:

admin. dist.	administrative district	for.	forest	pref.	prefecture
admin. div.	administrative division	g.	gulf	prov.	province
admin. reg.	administrative region	hd.	headland	pt	point
aut. reg.	autonomous region	i.	island	r.	river
aut. rep.	autonomous republic	imp. lake	impermanent lake	reg.	region
b.	bay	is	islands	resr	reservoir
c.	cape	l.	lake	salt l.	salt lake
depr.	depression	lag.	lagoon	sea chan.	sea channel
des.	desert	mt.	mountain	terr.	territory
disp terr.	disputed territory	mts	mountains	vol.	volcano
esc.	escarpment	pen.	peninsula		
est.	estuary	plat.	plateau		

1

9 de Julio 54D5
25 de Mayo 54D5

A

Aabenraa 11F9
Aachen 13K5
Aalborg 11F8
Aalborg Bugt b. 11G8
Aalen 13M6
Aalst 12J5
Aarhus 11G8
Aars 11F8
Aasiaat 45M3
Aba 32D4
Ābādān 33H1
Ābādeh 26E3
Abadla 32C1
Abaeté 55B2
Abaetetuba 53I4
Abakaliki 32D4
Abakan 24K4
Abakanskiy Khrebet mts24J4
Abancay 52D6
Abarkūh 26E3
Abashiri 30G3
Abbeville 18E1
Abbeville 47I6
Abéché 33F3
Abengourou 32C4
Abeokuta 32D4
Aberdare 15D7
Aberdeen 42E4
Aberdeen 16G3
Aberdeen 46H2
Abergavenny 15D7
Aberystwyth 15C6
Abhā 34E2
Abhar 33H1
Abidjan 32C4
Abilene 46H5
Abingdon 15F7
Abinsk 23H7
Abitibi, Lake 45J5
Aboisso 32C4
Abomey 32D4
Abong Mbang 32E4
Aboyne 16G3
Abqaiq 34E1
Abrantes 19B4
Absaroka Range mts 46E3
Abū 'Arīsh 34E2
Abu Dhabi 26E4
Abu Hamed 33G3
Abuja 32D4
Abū Kamāl 33H1
Abu Road 27G4
Açailândia 53I5
Acaponeta 50C4
Acapulco 50E5
Acará 53I4
Acaraú 53J4
Acarigua 52E2
Acatlán 50E5
Accra 32C4
Accrington 14E5
Achaguas 52E2
Acheng 30B3
Achinsk 24K4
Acıpayam 21M6
Acireale 20F6
Acklins Island 47M7
Acle 15I6
Aconcagua, Cerro mt. 54B4
Acopiara 53K5
A Coruña 19B2
Acquaviva delle Fonti 20G4
Acqui Terme 20C2
Acri 20G5
Ada 47H5
Adamantina 55A3
Adams 48E1
Adana 33G1
Adapazarı 21N4
Ad Dafinah 34E1
Ad Dahnā' des. 34E1
Ad Dār al Ḥamrā' 34D1
Ad Darb 34E2
Ad Dawādimī 34E1
Addis Ababa 34D3
Ad Dīwānīyah 33H1
Addlestone 15G7
Adelaide 41H6
Aden 34E2
Aden, Gulf of 34E2
Adigrat 34D2
Adīrī 33E2
Adirondack Mountains 48D1
Adjud 21L1
Admiralty Gulf 40F2
Admiralty Islands 38E2
Ado-Ekiti 32D4
Adrano 20F6
Adrar 32C2
Adrar, Dahr hills 32B3
Adrian 46G4
Adriatic Sea 20E2
Ādwa 34D2
Adzopé 32C4
Aegean Sea 21K5
A Estrada 19B2
Afanas'yevo 22L4
Afghanistan country 26F3
Afgooye 34E3
Afogados da Ingazeira 53K5
Afonso Cláudio 55C3
Afuá 53H4
Afyon 21N5
Agadez 32D3
Agadir 32C1
Agara 23I8
Agartala 27I4
Agboville 32C4
Ağcabädi 23J8
Agde 18F5
Agen 18E4
Agios Dimitrios 21J6
Agios Nikolaos 21K7
Agnibilékrou 32C4
Agnita 21K2
Agra 27G4
Ağrı 26D3
Agrigento 20E6
Agrinio 21I5
Aguadilla 51K5
Agua Prieta 46F5
Aguascalientes 50D4
Agudos 55A3
Águilas 19F5
Agulhas, Cape 36E8
Ahaggar plat. 32D2
Ahaggar, Tassili oua-n- plat. 32D2
Ahar 26D3
Ahmadabad 27G4
Ahmar Mountains mts 34E3
Ahtme 11O7
Ahvāz 33H1
Aigio 21J5
Aiken 47K5
Aïn Beïda 20B7
Aïn Defla 19H5
Aïn Deheb 19G6
Aïn el Hadjel 19H6
Aïn Oussera 19H6
Aïn Sefra 32C1
Aïn Taya 19H5
Aïn Tédélès 19G6
Aïn Temouchent 19F6
Aïr, Massif de l' mts 32D3
Airdrie 16F5
Aiud 21J1
Aix-en-Provence 18G5
Aix-les-Bains 18G4
Aizawl 27I4
Aizkraukle 11N8
Aizuwakamatsu 31E5
Ajaccio 18I6
Ajdābiyā 33F1
Ajmer 27G4
Akadyr 27G2
Akçakoca 21N4
Akchâr reg. 32B3
Akersberga 11K7
Aketi 34C3
Akhali Atoni 23I8
Akhḍar, Al Jabal al mts 33F1
Akhisar 21L5
Akhtubinsk 23J6
Akhty 23J8
Aki 31D6
Akita 31F5
Akjoujt 32B3
Akkajaure l. 10J3
Akkol' 27G1
Akkuş 23H8
Akom II 32E4
Akonolinga 32E4
Akordat 33G3
Akranes 10□
Åkrehamn 11D7
Akron 48A2
Aksai Chin disp. terr. 27G3
Aksay 23H7
Akşehir 21N5
Aksu 27H2
Aksubayevo 23K5
Aktau 26E2
Aktobe 26E1
Aktsyabrski 23F5
Akune 31C6
Akure 32D4
Akureyri 10□
Akwanga 32D4
Akyazı 21N4
Alabama r. 47J5
Alabama state 47J5
Alaçam 23G8
Alagir 23J8
Alagoinhas 55D1
Al Aḥmadī 26D4
Alajärvi 10M5
Al 'Alayyah 34E2
Al 'Amārah 33H1
Alameda 49A2
Al 'Āmirīyah 33F1
Alamogordo 46F5
Alamos 46F6
Alamosa 46F4
Åland Islands 11K6
Alanya 33G1
Alaplı 21N4
Al 'Aqabah 33G2
Al 'Aqīq 34E1
Al 'Arīsh 33G1
Al Arṭāwīyah 34E1
Alaşehir 21M5
Alaska state 44D3
Alaska, Gulf of 44D4
Alaska Peninsula 44B4
Alaska Range mts 44D3
Alatyr' 23J5
Alausí 52C4
Alaverdi 23J8
Alavus 10M5
Alba 20C2
Albacete 19F4
Alba Iulia 21J1
Albania country 21H4
Albany 40D6
Albany r. 45J4
Albany 47K5
Albany 48E1
Albany 46C3
Al Bayḍā' 33F1
Al Bayḍā' 34E2
Albenga 20C2
Albert, Lake 34D3
Alberta prov. 44G4
Albert Lea 47I3
Albert Nile r. 33G4
Alberton 37I4
Albertville 18H4
Albi 18F5
Albina 53H2
Albion 48B1
Al Birk 34E2
Al Biyāḍh reg. 34E1
Alborán, Isla de i. 19E6
Alboran Sea 19D5
Albufeira 19B5
Albuquerque 46F4
Al Buraymī 26E4
Albury 42C6
Alcalá de Henares 19E3
Alcalá la Real 19E5
Alcamo 20E6
Alcántara 19C4
Alcaraz 19E4
Alcázar de San Juan 19E4
Alchevs'k 23H6
Alcobaça 55D2
Alcoy-Alcoi 19F4
Alcúdia 19H4
Aldabra Islands 35E4
Aldan 25N4
Aldan r. 25N3
Aldeburgh 15I6
Aldershot 15G7
Aldridge 15F6
Aleg 32B3
Alegre 55C3
Alegrete 54E3
Aleksandrov 22H4
Aleksandrov Gay 23K6
Aleksandrovskoye 23I7
Aleksandrovsk-Sakhalinskiy 30F2
Aleksandry, Zemlya i. 24F1
Alekseyevka 23H6
Alekseyevka 23H6
Alekseyevskoye 22K5
Aleksin 23H5
Aleksinac 21I3
Alençon 18E2
Alenquer 53H4
'Alenuihāhā Channel 46□
Aleppo 33G1
Alès 18G4
Aleşd 21J1
Alessandria 20C2
Ålesund 10E5
Aleutian Islands 44A4
Aleutian Range mts 44C4
Alexander Bay 36C5
Alexandra 42B6
Alexandria 33F1
Alexandria 21K3
Alexandria 16E5
Alexandria 47I5
Alexandria 48C3
Alexandroupoli 21K4
Aleysk 24J4
Al Fayyūm 33G2
Alfenas 55B3
Alford 14H5
Alfred 48C1
Algarve reg. 19B5
Algeciras 19D5
Algemesí 19F4
Algeria country 32C2
Al Ghaydah 34F2
Alghero 20C4
Algiers 19H5
Algoa Bay 37G7
Algona 47I3
Algorta 19E2
Al Ḥadīthah 33H1
Alhama de Murcia 19F5
Al Ḥanākīyah 34E1
Al Ḥasakah 33H1
Al Ḥayy 33H1
Al Ḥinnāh 34E1
Al Hoceima 19E6
Al Hufūf 34E1
Aliağa 21L5
Alicante 19F4
Alice 46H6
Alice Springs 40G4
Alihe 30A2
Alindao 34C3
Alingsås 11H8
Al Ismā'īlīyah 33G1
Aliveri 21K5
Al Jahrah 26D4
Al Jawf 33F2
Al Jufrah 33E2
Al Jumaylīyah 34F1
Aljustrel 19B5
Al Kahfah 34E1
Al Khārijah 33G2
Al Khaṣab 26E4
Al Khawr 34F1
Al Khums 33E1
Al Khunn 34E1
Alkmaar 12J4
Al Kūt 33H1
Allahabad 27H4
Allakh-Yun' 25O3
Allegheny r. 48B2
Allegheny Mountains 48A4
Allende 46G6
Allentown 48D2
Alliance 46G3
Alliance 48A2
Allier r. 18F3
Al Līth 34E1
Alloa 16F4
Alma 47M2
Alma 19B4
Al Maḥwīt 34E2
Almansa 19F4
Al Manṣūrah 33G1
Al Marj 33F1
Almaty 27G2
Almaznyy 25M3
Almeirim 53H4
Almeirim 19B4
Almelo 13K4
Almenara 55C2
Almendralejo 19C4
Almería 19E5
Almería, Golfo de b. 19E5
Al'met'yevsk 24G4
Älmhult 11I8
Al Mindak 34E1
Al Minyā 33G2
Al Mubarrez 34E1
Alofi 39J3
Alor Star 29C7
Alotau 41K2
Alpena 47K2
Alpine 46G5
Alps mts 18H4
Al Qā'īyah 34E1
Al Qāmishlī 33H1
Al Qaṭn 34E2
Al Qunfidhah 34E2
Al Quṣayr 33G2
Al Quwayīyah 34E1
Alsager 15E5
Alstonville 42F2
Alta 10M2
Alta Floresta 53G5
Altai Mountains 27H2
Altamira 53H4
Altamura 20G4
Altay 27H2
Altay 27I2
Altdorf 18I3
Altıntaş 21N5
Altiplano plain 52E7
Alto Garças 53H7
Altoona 48B2
Alto Parnaíba 53I5
Altrincham 14E5
Altun Shan mts 27H3
Altus 46H5
Alūksne 11O8
Alva 46H4
Alvesta 11I8
Älvsbyn 10L4
Al Wajh 34D1
Al Widyān plat. 33H1
Alyangula 41H2
Alyth 16F4
Alytus 11N9
Amadeus, Lake salt flat 40G4
Amadora 19B4
Åmål 11H7
Amambaí 54E2
Amarante 53J5
Amareleja 19C4
Amargosa 55D1
Amarillo 46G4
Amasra 23G8
Amazar 30A1
Amazon r. 52F4
Amazon, Mouths of the 53I3
Ambalavao 35E6
Ambam 34B3
Ambato 52C4
Ambato Boeny 35E5
Ambato Finandrahana 35E6
Ambatolampy 35E5
Ambatondrazaka 35E5
Amberg 13M6
Ambilobe 35E5
Ambleside 14E4
Amboasary 35E6
Ambodifotatra 35E5
Ambohimahasoa 35E6
Ambon 29E8
Ambositra 35E6
Ambovombe 35E6
Ambriz 35B4
Americana 55B3
American Fork 46E3
American Samoa terr. 39J3
Americus 47K5
Amersfoort 12J4
Amersham 15G7
Ames 47I3
Amesbury 48F1
Amfissa 21J5
Amga 25O3
Amherst 48E1
Amiens 18F2
Amistad Reservoir 46G6
Amlwch 14C5
'Ammān 33G1
Ammanford 15D7
Ämmänsaari 10P4
Amorgos i. 21K6
Amos 45K5
Ampanihy 35E6
Amparo 55B3
Ampasimanolotra 35E5
Amravati 27G4
Amritsar 27G3
Amstelveen 12J4
Amsterdam 12J4
Amsterdam 48D1
Amstetten 13O6
Am Timan 33F3
Amudar'ya r. 26E2
Amundsen Gulf 44F2
Amundsen Sea 56C4
Amuntai 29D8
Amur r. 30D2
Amur r. 30F1
Amursk 30E2
Amvrosiyivka 23H7
Anabanua 29E8
Anaconda 46E2
Anadolu Dağları mts 26C2
Anadyr' 25S3
Anagé 55C1
'Anah 33H1
Anaheim 49D4
Anajás 53I4
Analalava 35E5
Anamur 33G1
Anan 31D6
Anantapur 27G5
Anan'yiv 23F7
Anapa 23H7
Anápolis 55A2
Añatuya 54D3
Anbyon 31B5
Anchorage 44D3
Ancona 20E3
Anda 30B3
Andacollo 54B4
Andalucía aut. comm. 19D5
Andaman Islands 27I5
Andaman Sea 29B6
Andapa 35E5
Andenne 12J5
Andéramboukane 32D3
Anderlecht 12J5
Anderson 47J3
Anderson 47K5
Andes mts 54C4
Andijon 27G2
Andilamena 35E5
Andilanatoby 35E5
Andoany 35E5
Andong 31C5
Andorra country 19G2
Andorra la Vella 19G2
Andover 15F7
Andradina 55A3
Andreapol' 22G4
Andrelândia 55B3
Andrews 46G5
Andria 20G4
Andros i. 47L7
Andros i. 21K6
Andselv 10K2
Andújar 19D4
Andulo 35B5
Anegada, Bahía b. 54D6
Aného 32D4
Ang'angxi 30A3
Angara r. 25K4
Angarsk 25L4
Angatuba 55A3
Angel Falls 52F2
Ängelholm 11H8
Angers 18D3
Anglesey i. 14C5
Angoche 35D5
Angol 54B5
Angola country 35B5
Angola 48B1
Angoulême 18E4
Angra dos Reis 55B3
Angren 27G2
Anguang 30A3
Anguilla terr. 51L5
Ankara 26C3
Ankazoabo 35E6
Anna 23I6
Annaba 20B6
An Nafūd des. 26D4
An Najaf 33H1
Annandale 48C3
Annapolis 48C3
Ann Arbor 47K3
Anna Regina 53G2
An Nāṣirīyah 33H1
Annecy 18H4
An Nimāṣ 34E2
Anniston 47J5
Ansbach 13M6
Anshan 30A4
Anshun 27J4
Antakya 33G1
Antalaha 35F5
Antalya 21N6
Antalya Körfezi g. 21N6
Antananarivo 35E5
Antarctica 56
Antarctic Peninsula 56D4
Antequera 19D5
Anti-Atlas mts 32C2
Antibes 18H5
Anticosti, Île d' i. 45L5
Antigua and Barbuda country 51L5
Antikythiro, Steno sea chan. 21J7
Antioch 49B1
Antipodes Islands 39H6
Antofagasta 54B2
Antrim 17F3
Antrim Hills 17F2
Antsalova 35E5
Antsirabe 35E5
Antsirañana 35E5
Antsohihy 35E5
Antwerp 12J5
Anuchino 30D4
Anuradhapura 27H6
Anxious Bay 40G6
Anyang 27K3
Anyang 31B5
Anzhero-Sudzhensk 24J4
Anzio 20E4
Aomori 30F4
Aoraki/Mount Cook mt. 43C6
Aosta 20B2
Aoukâr reg. 32C2
Aparecida do Tabuado 55A3
Apatity 10R3
Apatzingán 50D5
Apeldoorn 13J4
Apennines mts 20C2
Apia 39I3
Apiaí 55A4
Aporé 55A2
Appalachian Mountains 47K4
Appennino Abruzzese mts 20E3
Appennino Tosco-Emiliano mts 20D3
Appennino Umbro-Marchigiano mts 20E3
Appleton 47J3
Apple Valley 49D3
Aprilia 20E4
Apsheronsk 23H7
Apt 18G5
Apucarana 55A3
Aqaba, Gulf of 26C4
Aquidauana 54E2
Arabian Peninsula 26D4
Arabian Sea 26F5
Aracaju 53K6
Aracati 53K4
Aracatu 55C1
Araçatuba 55A3
Aracruz 55C2
Araçuaí 55C2
Arad 21I1
Arafura Sea 38D2
Aragarças 53H7
Araguaçu 55A1
Araguaína 53I5
Araguari 55A2
Araguatins 53I5
Araioses 53J4
Arāk 33H1
Arakan Yoma mts 27I4
Aral Sea salt l. 26F2
Aral'sk 26F2
Aranda de Duero 19E3
Aranđelovac 21I2
Aranjuez 19E3
Arao 31C6
Arapiraca 53K5
Arapongas 55A3
Araquari 55A4
'Ar'ar 26D3
Araranguá 55A5
Araraquara 55A3
Araripina 53J5
Arataca 55D1
Arauca 52D2
Arawa 38F2
Araxá 55B2
Arayıt Dağı mt. 21N5
Arbil/Hewlêr 33H1
Arboga 11I7
Arbroath 16G4
Arcachon 18D4
Arcade 48B1
Arcadia 47K6
Arcelia 50D5
Archangel 22I2
Arcos 55B3
Arcos de la Frontera 19D5
Ardabīl 26D3
Ardahan 23I8
Ardatov 23I5
Ardatov 23J5
Ardee 17F4
Ardennes plat. 12J6
Arden Town 49B1
Ardestān 26E3
Ardmore 47H5
Ardrossan 16E5
Areia Branca 53K4
Arendal 11F7
Arequipa 52D7
Arezzo 20D3
Arganda del Rey 19E3
Argentan 18D2
Argentina country 54C5
Argentino, Lago l. 54B8
Argos 21J6
Argostoli 21I5
Argun' r. 28E2
Argun 23J8
Argungu 32D3
Argyle, Lake 40F3
Ariano Irpino 20F4
Aribinda 32C3
Arica 52D7
Arima 51L6
Arinos 55B1
Aripuanã 53G6
Ariquemes 52F5
Arisaig, Sound of sea chan. 16D4
Arizona state 46E5
'Arjah 34E1
Arkadak 23I6
Arkadelphia 47I5
Arkalyk 26F1
Arkansas r. 47I5
Arkansas state 47I4
Arkansas City 47H4
Arkhara 30C2
Arklow 17F5
Arkul' 22K4
Arles 18G5
Arlington 48E2
Arlington 48C3
Arlon 13J6
Armagh 17F3
Armant 33G2
Armavir 23I7
Armenia country 26D2
Armenia 52C3
Armeria 50D5
Armidale 42E3
Arnhem 13J5
Arnhem Land reg. 40G2
Arnold 15F5
Arnsberg 13L5
Arquipélago da Madeira aut. reg. 32B1
Arraias 55B1
Ar Ramādī 33H1
Arran i. 16D5
Ar Raqqah 33G1
Arras 18F1
Ar Rayyān 34F1
Arrecife 32B2
Arriagá 50F5
Ar Rifā'ī 33H1
Arroyo Grande 49B3
Arsen'yev 30D3
Arsk 22K4
Arta 21I5
Artem 30D4
Artemivs'k 23H6
Artesia 46G5
Artigas 54E4
Art'ik 23I8
Artsyz 21M2
Artvin 23I8
Arua 34D3
Aruba terr. 51K6
Arundel 15G8
Arusha 34D4
Arvayheer 27J2
Arviat 45I3
Arvidsjaur 10K4
Arvika 11H7
Arzamas 23I5
Arzew 19F6
Arzgir 23J7
Asaba 32D4
Asahikawa 30F4
Asalē l. 34E2
Asansol 27H4
Āsayita 34E2
Asbury Park 48D2
Ascensión 52F7
Ascension i. 6
Aschaffenburg 13L6
Ascoli Piceno 20E3
Åseda 11I8
Asenovgrad 21K3
Asheville 47K4
Ashgabat 26E3
Ashford 15H7
Ashibetsu 30F4
Ashikaga 31E5
Ashington 14F3
Ashland 47K4
Ashland 47I2
Ashmore and Cartier Islands terr. 40E2
Ashmyany 11N9
Ash Shiḥr 34E2
Ashtabula 48A2
Ashton-under-Lyne 14E5
Asilah 19C6
Asinara, Golfo dell' b. 20C4
Asino 24J4
Asipovichy 23F5
'Asīr reg. 34E1
Asker 11G7
Askim 11G7
Asmara 33G3
Āsosa 34D2
Aspatria 14D4
Aspen 46F4
Assab 33H3
As Samāwah 33H1
Assen 13K4
Assiniboine r. 44I5
Assis 55A3
Assisi 20E3
As Sulaymānīyah/Slêmânî 33H1
Assynt, Loch l. 16D2
Astakos 21I5
Astana 27G1
Āstārā 26D3
Asti 20C2
Astorga 19C2
Astoria 46C2
Åstorp 11H8
Astrakhan' 23K7
Astravyets 11N9
Asturias aut. comm. 19C2
Asunción 54E3
Aswān 33G2
Asyūṭ 33G2
Atacama, Salar de salt flat 54C2
Atacama Desert 54C3
Atakpamé 32D4
Ataléia 55C2
Atamyrat 26F3
'Ataq 34E2
Atâr 32B2
Atascadero 49B3
Atasu 27G2
Atbara 33G3
Atbara r. 33G3
Atbasar 26F1
Atchison 47H4
Athabasca, Lake 44H4
Athens 21J6
Athens 47J5
Athens 47K4
Athens 48C2
Athens 47K4
Atherstone 15F6
Athlone 17E4
Athol 48E1
Athy 17F5
Ati 33E3
Atico 52D7
Atka 25Q3
Atkarsk 23J6
Atlanta 47K5
Atlantic 47H3
Atlantic City 48D3
Atlantis 36D7
Atlas Mountains 32C1
Atlas Saharien mts 32D1
Atlas Tellien mts 19H6
Aṭ Ṭā'if 34E1
Attu Island 25S4
Aṭ Ṭūr 33G2
Åtvidaberg 11I7
Atwater 49B2
Atyashevo 23J5
Atyrau 26E2
Atyrauskaya Oblast' admin. div. 23K7
Aubagne 18G5
Auburn 49B1
Auburn 48C1
Auch 18E5
Auchterarder 16F4
Auckland 43E3
Auckland Islands 39G7
Audo Range mts 34E3
Augsburg 13M6
Augusta 40D6
Augusta 20F6
Augusta 47K5
Augusta 47N3
Auki 41M1
Aurangabad 27G5
Aurich 13K4
Aurillac 18F4
Aurora 46G4
Aurora 47J3
Austin 47I3
Austin 46H5
Austintown 48A2
Australia country 40F5
Australian Capital Territory admin. div. 42D5
Austria country 13N7
Autazes 53G4
Auvergne, Monts d' mts 18F4
Auxerre 18F3
Avaré 55A3
Aveiro 19B3
Avellino 20F4
Aversa 20F4
Avesta 11J6
Avezzano 20E3
Aviemore 16F3
Avignon 18G5
Ávila 19D3
Avilés 19D2
Avoca 42A6
Avola 20F6
Avon r. 15F8
Avon r. 16F3
Avon 48C1
Awbārī 32E2
Awe, Loch l. 16D4
Aweil 33F4
Awka 32D4
Axminster 15E8
Ayacucho 54E5
Ayacucho 52D6
Ayagoz 27H2
Ayancık 23G8
Ayang 31B5
Aydın 21L6
Aylesbury 15G7
'Ayoûn el 'Atroûs 32C3
Ayr 16E5
Aytos 21L3
Ayutthaya 29C6
Ayvacık 21L5
Ayvalık 21L5
Azaouâd reg. 32C3
Azare 32E3
Azawad reg. 32C3
Azerbaijan country 26D2
Azogues 52C4
Azores terr. 6
Azov 23H7
Azov, Sea of 23H7
Azzaba 20B6
Az Zaqāzīq 33G1
Az Zarqā' 33G1
Azzeffāl hills 32B2

B

Baardheere 34E3
Babadag 21M2
Babaeski 21L4
Babahoyo 52C4
Bāb al Mandab strait 34E2
Babanusa 33F3
Babati 35D4
Babayevo 22G4
Babayurt 23J8
Babruysk 23F5
Bacabal 53J4
Bacău 21L1
Bačka Palanka 21H2
Bac Liêu 29C7
Bacolod 29E6
Badajoz 19C4
Baden 13P6
Baden 18I3
Baden-Baden 13L6
Bad Hersfeld 13L5
Bad Ischl 13N7
Bad Kissingen 13M5
Badou 32D4
Badr Ḥunayn 34D1
Bad Salzungen 13M5
Bad Schwartau 13M4
Badulla 27H6
Bafatá 32B3
Baffin Bay sea 45L2
Baffin Island 45L3
Bafia 32E4
Bafilo 32D4
Bafoussam 32E4
Bāfq 26E3
Bafra 23G8
Bāft 26E4
Bafwasende 34C3
Bagamoyo 35D4
Bagé 54F4
Bagenalstown 17F5
Baghdād 33H1
Baghlān 27F3
Bagrationovsk 11L9
Bahia state 55C1
Bahía Blanca 54D5
Bahir Dar 34D2
Bahrain country 34F1
Bahrīyah, Wāḥāt al oasis 33F2
Baia Mare 21J1
Baicheng 30A3
Baie-Comeau 45L5
Baikal, Lake 25L4
Băileşti 21J2
Bainbridge 47K5
Baiquan 30B3
Bairnsdale 42C6
Baise 27J4
Baishan 30B4
Baisogala 11M9
Baiyuda Desert 33G3
Baja 20H1
Baja California pen. 46D5
Bajoga 32E3
Bakel 32B3
Baker City 46D3
Baker Island terr. 39I1
Baker Lake 45I3
Baker Lake l. 45I3
Bakersfield 49C3
Bakhmach 23G6
Bakırköy 21M4
Baksan 23I8
Baku 26D2
Balabac Strait strait 29D7
Balakän 23J8
Balakhna 22I4
Balakliya 23H6
Balakovo 23J5
Balanga 29E6
Balashov 23I6
Balaton, Lake 20G1
Balatonboglár 20G1
Balatonfüred 20G1
Balbina, Represa de resr 53G4
Balchik 21M3
Baldwin 48B2
Balde 32B3
Balearic Islands is 19G4
Baleshwar 27H4
Baléyara 32D3
Bali i. 29D8
Bali, Laut sea 29D8
Balige 27I6
Balıkesir 21L5
Balikpapan 29D8
Balimo 38E2
Balingen 13L6
Balkanabat 26E3
Balkan Mountains 21J3
Balkash 27G2
Balkhash, Lake 27G2
Ballarat 42A6
Ballari 27G5
Ballé 32C3
Ballina 42F2
Ballina 17C3
Ballinasloe 17D4
Ballyclare 17G3
Ballymena 17F3
Ballymoney 17F2
Ballynahinch 17G3
Ballyshannon 17D3
Balranald 42A5
Balş 21K2
Balsas 53I5
Balta 23F7
Bălţi 23E7
Baltic Sea g. 11J9
Baltimore 48C3
Baltiysk 11K9
Balvi 11O8
Balykchy 27G2
Bam 26E4
Bamako 32C3
Bamba 32C3
Bambari 34C3
Bamberg 13M6
Bambuí 55B3
Bamenda 32E4
Banamba 32C3
Banaz 21M5
Banbridge 17F3
Banbury 15F6
Banda, Laut sea 29F8
Banda Aceh 27I6
Bandar-e 'Abbās 26E4
Bandar-e Būshehr 26E4
Bandar-e Jāsk 26E4
Bandar-e Kangān 26E4
Bandar-e Lengeh 26E4
Bandar Lampung 29C8
Bandar Seri Begawan 29D7
Bandiagara 32C3
Bandırma 21L4
Bandon 17D6
Bandundu 34B4
Bandung 29C8
Banes 51I4
Banff 44G4
Banff 16G3
Banfora 32C3
Bangassou 34C3
Bangka i. 29C8
Bangko 29C8
Bangkok 29C6
Bangladesh country 27H4
Bangolo 32C4
Bangor 17G3
Bangor 14C5
Bangor 47N3
Bangui 34B3
Banī Suwayf 33G2
Banī Walīd 33E1
Banja Luka 20G2
Banjarmasin 29D8
Banjul 32B3
Bankilaré 32D3
Banks Peninsula 43D6
Bann r. 17F2
Banning 49D4
Banská Bystrica 13Q6
Bantry 17C6
Bantry Bay 17C6
Banyo 32E4
Baochang 27K2
Baoding 27K3
Baoji 27J3
Baoqing 30D3
Baoshan 27I4
Baotou 27K2
Ba'qūbah 33H1
Bar 21H3
Baracoa 51J4
Barahona 51J5
Barakaldo 19E2
Baranavichy 11O10
Baranīs 33G2
Baraouéli 32C3
Barbacena 55C3
Barbados country 51M6
Barbate 19D5
Barcaldine 41J4
Barcelona 19H3
Barcelona 52F1
Barcelos 52F4
Barclayville 32C4
Barcs 20G2
Bärdä 23J8
Bardejov 23D6
Bareilly 27G4
Barents Sea 24F2
Barentu 33G3
Barham 42B5
Bari 20G4
Barinas 52D2
Bariri 55A3
Barisan, Pegunungan mts 29C8
Barkly East 37H6
Barkly Tableland reg. 41H3
Barkly West 36G5
Barkol 27I2
Bârlad 21L1
Bar-le-Duc 18G2
Barlee, Lake salt flat 40D5
Barletta 20G4
Barmer 27G4
Barmouth 15C6
Barnaul 24J4
Barnsley 14F5
Barnstaple 15C7
Barnstaple Bay 15C7
Baro 32D4
Barquisimeto 52E1
Barra 53J6
Barra i. 16B4
Barra, Sound of sea chan. 16B3
Barraba 42E3
Barra Bonita 55A3
Barra do Bugres 53G7
Barra do Corda 53I5
Barra do Garças 53H7
Barra do Piraí 55C3
Barra Mansa 55B3
Barranqueras 54E3
Barranquilla 52D1
Barreiras 53J6
Barreirinha 53G4
Barreirinhas 53J4
Barreiro 19B4
Barreiros 53K5
Barretos 55A3
Barrhead 16E5
Barrier Range hills 41I6
Barro Alto 55A1
Barrow-in-Furness 14D4
Barry 15D7
Barstow 49D3
Bartica 53G2
Bartın 23G8
Bartoszyce 13R3
Baruun-Urt 27K2
Barwon r. 42C3
Barysaw 11P9
Barysh 23J5
Basarabi 21M2
Basel 18H3
Bashmakovo 23I5
Bashtanka 23G7
Basildon 15H7
Basingstoke 15F7
Basra 33H1
Bassano del Grappa 20D2
Bassar 32D4
Bassein 27I5
Basse-Normandie admin. reg. 15F9
Basse Santa Su 32B3
Basse-Terre 51L5
Basseterre 51L5
Bass Strait strait 42C7
Bastia 18I5
Bastogne 13J5
Bastrop 47I5
Bata 32D4
Batabanó, Golfo de b. 51H4
Batagay 25O3
Batangafo 34B3
Batangas 29E6
Batavia 48B1
Bataysk 23H7
Bătdâmbâng 29C6
Batemans Bay 42E5
Batesville 47I4
Bath 15E7
Bathgate 16F5
Bathurst 42D4
Bathurst 47N2
Bathurst Inlet 44H3
Bathurst Island 40G2
Bathurst Island 45I2
Batié 32C4
Batı Menteşe Dağları mts 21L6
Batı Toroslar mts 21N6
Batley 14F5
Batlow 42D5
Batna 32D1
Baton Rouge 47I5
Batouri 33E4
Batticaloa 27H6
Battipaglia 20F4
Battle Creek 47J3
Bat'umi 23I8
Baturité 53K4
Batyrevo 23J5
Baubau 29E8
Bauchi 32D3
Bauru 55A3
Bauska 11N8
Bautzen 13O5
Bayamo 51I4
Bayan 30B3
Bayanhongor 27J2
Bayan Hot 27J3
Bayannur 27J2
Bay City 47K3
Bay City 47H6
Baydhabo 34E3
Bayerischer Wald mts 13N6
Bayeux 15G9
Bayındır 21L5
Baykonyr 26F2
Baymak 24G4
Bayombong 29E6
Bayonne 18D5
Bayonne 48D2
Bayramiç 21L5
Bayreuth 13M6
Bay Shore 48E2
Baza 19E5
Bazarnyy Karabulak 23J5
Beachy Head 15H8
Beacon 48E2
Beacon Bay 37H7
Beaconsfield 15G7
Beagle Gulf 40F2
Bealanana 35E5
Beaminster 15E8
Beatrice 47H3
Beaufort 42A6
Beaufort Sea 44D2
Beaufort West 36F7
Beaumont 47I5
Beaune 18G3
Beauvais 18F2
Beaver 48A2
Beaver Falls 48A2
Beawar 27G4
Bebedouro 55A3
Bebington 14D5
Beccles 15I6
Bečej 21I2
Becerreá 19C2
Béchar 32C1
Beckley 48A4
Bedale 14F4
Bedford 37J5
Bedford 15G6
Bedford 47J4
Bedford 48B2
Bedford 48B4
Bedlington 14F3
Bedworth 15F6
Beechworth 42C6
Beenleigh 42F1
Beersheba 33G1
Beeville 46H6
Bega 42D6
Behshahr 26E3
Bei'an 30B2
Beihai 27J4
Beijing 27K3
Beira 35D5
Beirut 33G1
Beja 19C4
Béja 20C6
Bejaïa 19I5
Békés 21I1
Békéscsaba 21I1
Bekovo 23I5
Bela 26F4
Bela-Bela 37I3
Bélabo 32E4
Belagavi 27G5
Bel Air 48C3
Belalcázar 19D4
Belarus country 23E5
Bela Vista 54E2
Bela Vista de Goiás 55A2
Belaya Glina 23I7
Belaya Kalitva 23I6
Belaya Kholunitsa 22K4
Bełchatów 13Q5
Beledweyne 34E3
Belém 53I4
Belev 23H5
Belfast 17G3
Belfast 47N3
Belfast Lough inlet 17G3
Belfort 18H3
Belgium country 12J5
Belgorod 23H6
Belgrade 21I2
Belinskiy 23I5
Belinyu 29C8
Belize 35B4
Belize 50G5
Belize country 50G5
Bella Unión 54E4
Belledonne mts 18G4
Bellefonte 48C2
Belle Glade 47K6
Belle Isle, Strait of 45M4
Bellevue 46C2
Bellingham 46C2
Bellinzona 18I3
Belluno 20E1
Bell Ville 54D4
Bellville 36D7
Belmont 42E4
Belmonte 55D1
Belmopan 50G5
Belo Campo 55C1
Belogorsk 30C2
Belo Horizonte 55C2
Beloit 47J3
Belomorsk 22G2
Beloretsk 24G4
Belo Tsiribihina 35E5
Beloye, Ozero l. 22H3
Belozersk 22H3
Belyy 22G5
Bemidji 47I2
Ben Arous 20D6
Benavente 19D2
Benbecula i. 16B3
Bend 46C3
Bender 21M1
Bendigo 42B6
Benešov 13O6
Benevento 20F4
Bengal, Bay of sea 27H5
Bengaluru 27G5
Bengbu 27K3
Benghazi 33F1
Bengkulu 29C8
Benguela 35B5
Beni 34C3
Benidorm 19F4
Beni Mellal 32C1
Benin country 32D4
Benin, Bight of g. 32D4
Benin City 32D4
Beni Saf 19F6
Benito Juárez 54E5
Benjamin Constant 52E4
Ben Nevis mt. 16D4
Bennington 48E1
Benoni 37I4
Bentiu 33F4
Bento Gonçalves 55A5
Benton Harbor 47J3
Bentonville 47I4
Benue r. 32D4
Benxi 30A4
Béoumi 32C4
Beppu 31C6
Berat 21H4
Berber 33G3
Berbera 34E2
Berbérati 34B3
Berdyans'k 23H7
Berdychiv 23F6
Berehove 23D6
Berekum 32C4
Berezivka 23F7
Berezne 23E6
Bereznik 22I3
Berezniki 24G4
Berezovo 24H3
Bergama 21L5
Bergamo 20C2
Bergen 11D6
Bergerac 18E4
Bergheim 13K5
Bergsviken 10L4
Beringovskiy 25S3
Bering Sea 25S4
Bering Strait strait 44B3
Berkane 19E6
Berkeley 49A2
Berkovitsa 21J3
Berlin 13N4
Bermagui 42E6
Bermejo 52F8
Bermuda terr. 51L2
Bern 18H3
Bernardino de Campos 55A3
Berner Alpen mts 18H3
Beroun 13O6
Berriane 32D1
Berrouaghia 19H5
Berry 42E5
Bertolínia 53J5
Bertoua 32E4
Beruri 52F4
Berwick-upon-Tweed 14E3
Beryslav 21O1
Besalampy 35E5
Besançon 18H3
Beslan 23J8
Bessbrook 17F3
Bessemer 47J5
Bessonovka 23J5
Bethel Park 48A2
Bethesda 14C5
Bethesda 48C3
Bethlehem 37I5
Bethlehem 48D2
Betim 55B2
Betpakdala plain 26G2
Betroka 35E6
Bettiah 27H4
Bettystown 17F4
Beverley 14G5
Beverly 48F1

Beverly Hills 49C3
Bexhill 15H8
Beykoz 21M4
Beyla 32C4
Beyneu 26E2
Beypazarı 21N4
Beyşehir 33G1
Bezhanitsy 22F4
Bezhetsk 22H4
Béziers 18F5
Bhamo 27I4
Bhavnagar 27G4
Bhekuzulu 37J4
Bhilwara 27G4
Bhiwani 37H7
Bhopal 27G4
Bhubaneshwar 27H4
Bhuj 27F4
Bhutan country 27I4
Biała Podlaska 23D5
Białogard 13D5
Biarritz 18D5
Bibai 30F4
Biberach an der Riß 13L6
Bicas 55C3
Bicester 15F7
Bida 32D4
Biddeford 48E2
Bideford 15C7
Bié, Planalto do 35B5
Bielawa 13P5
Biel/Bienne 18H3
Bielefeld 13L4
Biella 20C2
Bielsko-Biała 13Q6
Biên Hoa 29C6
Biga 21L4
Bigadiç 21M5
Biga Yarımadası pen. 21L5
Biggar 46F1
Biggleswade 15G6
Bighorn Mountains 46F3
Bignona 32B3
Big Rapids 47J3
Big Spring 46G5
Big Trout Lake 45I4
Bihać 20F2
Bijār 33I1
Bijeljina 21H2
Bijelo Polje 21H3
Bikaner 27G4
Bikin 30D3
Bila Tserkva 23F6
Bilbao 19E2
Bilecik 21M4
Biłgoraj 23D6
Bilhorod-Dnistrovs'kyy 21N1
Bilibino 25R3
Billericay 15H7
Billingham 14F4
Billings 46F2
Bill of Portland hd 15E8
Bilma, Grand Erg de des. 32E3
Bilohirs'k 23G7
Bilohir''ya 23E6
Bilovods'k 23H6
Biloxi 47J5
Biltine 33F3
Bilyayivka 21N1
Bimini Islands 47L6
Bindura 35D2
Binghamton 48D1
Bintulu 29D7
Binxian 30B3
Bioko i. 32D4
Bira 34C2
Bireun 27I6
Birigui 55A3
Birjand 26E3
Birkenhead 14D5
Birkirkara 20F7
Birmingham 15F6
Birmingham 47J5
Birnin-Gwari 32D3
Birnin Konni 32D3
Birobidzhan 30D2
Biržai 11N8
Bisbee 46F5
Biscay, Bay of sea 18B4
Bishkek 27J2
Bishop Auckland 14F4
Bishop's Stortford 15H7
Biskra 32D1
Bismarck 46G2
Bismarck Archipelago is 38E2
Bismarck Sea 38E2
Bissau 32B3
Bistrița 21K1
Bitola 21I4
Bitonto 20G4
Bitterroot Range mts 46D2
Biu 32E3
Biwa-ko l. 31D6
Biysk 24J4
Bizerte 20C6
Bjästa 10K5
Bjelovar 20G2
Bjerringbro 11F8
Björklinge 11J6
Bjørnøya i. 24C2
Bla 32D3
Blackall 38E4
Blackburn 14E5
Black Forest mts 13L7
Blackpool 14D5
Blacksburg 48A4
Black Sea 23H8
Blackwater r. 17F3
Blagoevgrad 21J3
Blagoveshchensk 30B2
Blanc, Mont mt. 18H4
Blanca, Bahía b. 54D5
Blanche, Lake salt flat 41H5
Blanes 19H3
Blansko 13P6
Blantyre 35D5
Blayney 42D4
Blenheim 43D5
Blessington Lakes 17F4
Bletchley 15G6
Blida 19H5
Bloemfontein 37H5
Bloomington 37I5
Bloomington 47J4
Bloomsburg 48C2
Bloxham 15F6
Bluefield 48A4
Bluefields 51H4
Blue Mountains 42D4
Blue Nile r. 33G3
Blue Ridge 48A4
Blue Ridge mts 48A4
Blumenau 55A4
Blyth 14F3
Blytheville 47J4
Bo 32B4
Boa Esperança 55B3
Boa Nova 55D2
Boa Viagem 55B2
Boa Vista 52F3
Bobo-Dioulasso 32C3
Bobrov 23I6

Bobrovytsya 23F6
Bobrynets' 23G6
Boca do Acre 52C5
Bocaiúva 55C2
Bocaranga 34B3
Bocas del Toro 51H7
Bochnia 13R6
Bochum 13K5
Boda 34B3
Bodaybo 25M4
Bodmin 15C8
Bodmin Moor moorland 15C8
Bodø 10L4
Bodrum 21L6
Boende 33F5
Bogalusa 47J5
Bogandé 32C3
Boggeragh Mountains hills 17C5
Bognor Regis 15G8
Bogoroditsk 23I5
Bogorodsk 22I4
Bogotá 52D3
Boguchany 25K4
Boguchar 23I6
Bogué 32B3
Bo Hai b. 27K3
Bohai Wan b. 27K3
Bohlokong 37I5
Böhmer Wald mts 13N6
Bohodukhiv 23G6
Bohol Sea 29E7
Bohu 27H2
Boise 46D3
Bojnūrd 26E3
Boké 32B3
Bokovskaya 23I6
Bol 33E3
Bolama 32B3
Bolbec 15H9
Bole 27H2
Bolgar 23K5
Bolgatanga 32C3
Bolhrad 21M2
Bolí 10C3
Boliden 10L4
Bolintin-Vale 21K2
Bolívar 52C5
Bolivia country 52E7
Bolkhov 23I5
Bollnäs 11J6
Bollstabruk 10J5
Bolobo 34B4
Bologna 20D2
Bologoye 22G4
Bolsena 11I9
Bol'shaya Glushitsa 23K5
Bol'shaya Martynovka 23I7
Bol'shevik, Ostrov i. 25L2
Bol'shoy Kamen' 30D4
Bol'shoy Murashkino 22J5
Bolton 14E5
Bolu 21N4
Bolvadin 21N5
Bolzano 20D1
Boma 34B4
Bomaderry 42E5
Bombala 42D5
Bomdila 27I4
Bom Jardim de Goiás 55A2
Bom Jesus 55A5
Bom Jesus da Lapa 55C1
Bom Sucesso 55B3
Bon, Ras 48C4
Bonaire mun. 51J7
Bonaparte Archipelago is 40E2
Bondo 34C3
Bondoukou 32C4
Bone, Teluk b. 29E8
Bo'ness 16F4
Bonete, Cerro mt. 54C3
Bongaigaon 27I4
Bongandanga 34C3
Bongor 33E3
Bonifacio 18D3
Bonifacio, Strait of strait 18H6
Bonin Islands 31F8
Bonn 13K5
Bonneville 18H3
Bonnyrigg 16F5
Bontoc 29E6
Bontosunggu 38B2
Boonah 42F1
Boone 48B1
Booneville 47J5
Boorowa 42D5
Boothia, Gulf of 45J3
Boothia Peninsula 45I2
Bootle 14E5
Bor 22J4
Bor 21I2
Borås 11H8
Borāzjān 26E4
Borba 53G4
Borça 20I3
Bordeaux 18D4
Borden Island 45G2
Bordj Bou Arréridj 19I5
Borgarnes 10C2[c]
Borislav 23I6
Borisoglebsk 23I6
Borisovka 23H6
Borlänge 11I6
Borneo i. 29D7
Bornholm i. 11I9
Bornova 21L5
Borovan 32C3
Borovichi 22G4
Borovoy 10R4
Borovoy 22L3
Borşa 23E7
Borshchiv 23E6
Borūjerd 33H1
Boryslav 23I6
Boryspil' 23F6
Borzya 25M4
Bosanska Gradiška 18E4
Bosnia and Herzegovina country 20G2
Bosobolo 34B3
Bosporus strait 21M4
Bossangoa 34B3
Bossembélé 34B3
Bossier City 47I5
Boston 15G6
Boston 48E2
Botany Bay 42E4
Botevgrad 21I3
Bothnia, Gulf of 11K6
Botlikh 23J8
Botoşani 21M1
Botrange mt. 18H3
Botshabelo 37H5
Botswana country 35C6
Bottrop 13K5
Botucatu 55A3
Botuporã 55C1
Bouaflé 32C4
Bouaké 32C4
Bouar 34B3
Bou Arfa 32C1
Bouca 34B3
Bougaa 34B3

Bougainville Island 38F2
Bougouni 32C3
Bougtob 32D1
Bouira 19H5
Boujdour 32B2
Boulder 46F1
Boulder City 49E3
Boulogne-Billancourt 18F2
Boulogne-sur-Mer 15I8
Boumerdès 19H5
Boundiali 32C4
Boundji 34B4
Bounty Islands 39H6
Bourail 39G4
Bourg-Achard 15H9
Bourg-en-Bresse 18G3
Bourges 18F3
Bourke 42D3
Bournemouth 15F8
Bou Saâda 19I6
Boutilimit 32B3
Bovalino 20G6
Boyle 17D4
Bozcaada i. 21L5
Bozdoğan 21M6
Bozeman 46E2
Bozoum 34B3
Bozüyük 21N5
Bra 20B3
Brač i. 20G3
Bracknell 15G7
Bradenton 47K6
Brades 51L5
Bradford 14F5
Brady 46H5
Braga 19B3
Bragado 54D5
Bragança 55I4
Bragança 19C3
Bragança Paulista 55B3
Brahin 23F6
Brahmapur 27H5
Brahmaputra r. 28B5
Brăila 21L2
Brainerd 47I2
Braintree 15H7
Bramming 11F9
Brampton 14E4
Branco r. 52F4
Brandberg mt. 35B6
Brande 11F9
Brandenburg an der Havel 13N4
Brandon 45I5
Braniewo 13Q3
Brantford 48A1
Branxton 42E4
Brasil, Planalto do plat. 55C2
Brasiléia 52C6
Brasília 55B1
Brasília de Minas 55B2
Braslaw 11O9
Braşov 21K2
Bratsk 25L4
Braunau am Inn 13N6
Braunschweig 13M4
Bravo del Norte, Río r. 46H4
Brawley 49E4
Bray 17F4
Brazil country 53G5
Brazos r. 47H6
Brazzaville 33B4
Brčko 20H2
Břeclav 13P6
Brecon 15D7
Brecon Beacons reg. 15D7
Breda 12J5
Bredasdorp 36E8
Bregenz 13L7
Breiðafjörður b. 10C2[c]
Bremen 13L4
Bremerhaven 13L4
Brenham 47H5
Brenner Pass pass 20D1
Brentwood 15H7
Brescia 20D2
Bressuire 18D3
Brest 11M10
Brest 18B2
Breton Sound b. 47J6
Breves 53H4
Brewarrina 42C3
Brewster 48A2
Brezno 13Q6
Bria 34C3
Briançon 18H4
Bridgend 15D7
Bridgeport 48E2
Bridgeton 48D3
Bridgetown 51M6
Bridgnorth 15E6
Bridgwater 15D7
Bridgwater Bay 15D7
Bridlington 14G4
Bridlington Bay 14G4
Bridport 42E5
Bridport 15E8
Brig 18H3
Brigham City 46E3
Brighton 15G8
Brighton 48C1
Brignoles 18H5
Brikama 32B3
Brindisi 20G4
Brisbane 42F1
Bristol 15E7
Bristol 48E2
Bristol 48C1
Bristol 47K4
Bristol Bay 44B4
Bristol Channel est. 15C7
British Columbia prov. 44F4
British Indian Ocean Territory terr. 7
Brits 37H3
Britstown 36F6
Brive-la-Gaillarde 18E4
Brixham 15D8
Brno 13P6
Broadstairs 15I7
Brockton 48F2
Brodnica 13Q4
Brody 23E6
Broken Arrow 47H4
Broken Hill 41I6
Brokopondo 53G2
Brønderslev 11F7
Brookhaven 47I5
Brookings 46F3
Brookings 47H3
Brookline 48F1
Brooks Range mts 44D3
Broome 40E3
Brora r. 17E4
Brosna r. 17E4
Brovary 23F6
Brownfield 46G5
Brownhills 15E6
Brownsville 48B2
Brownsville 47H6

Brownwood 46H5
Bruay-la-Buissière 18F1
Bruck an der Mur 13O7
Brugge 12I5
Brumado 55C1
Brumunddal 11F6
Brunei country 29D7
Brunflo 10I5
Brunswick 47K5
Brunswick 47N3
Bruntál 13P6
Brunswick 55A4
Brussels 12I5
Bryan 47H5
Bryansk 23G5
Bryne 11D7
Buala 41L1
Bucak 21N6
Bucaramanga 52D2
Bucharest 21L2
Bucheon 31B5
Buchholz 48A3
Buckhaven 16F4
Buckie 16G3
Buckingham 15G6
Buckingham Bay 41H2
Buda-Kashalyova 23F5
Budapest 21H1
Bude 15C8
Budennovsk 23J7
Buderim 42F1
Buenaventura 52C3
Buenos Aires 54E4
Buerarema 55D1
Buffalo 48B1
Buftea 21K2
Bug r. 13S5
Buga 52C3
Bugt 30A2
Buhuşi 21L1
Builth Wells 15D6
Buinsk 23K5
Bujanovac 21I3
Bujumbura 34C4
Bukavu 34C4
Bukittinggi 29C8
Bukoba 34D4
Bûlach 18H3
Bulawayo 35C6
Buldan 21M5
Bulembu 37J3
Bulgan 27J2
Bulgaria country 21K3
Bulldhead City 49E3
Bumba 34C3
Bunbury 40D6
Bundaberg 41K4
Bundoran 17D3
Bungay 15I6
Bungendore 42E5
Bunia 34D3
Buôn Ma Thuột 29C6
Buôrkhaya 34E1
Burao 34E3
Burbank 49C3
Burco 34E3
Burdur 21N6
Bure 34D2
Bureá 10L4
Burgas 21L3
Burgaw 45M5
Burgersdorp 37H6
Burgess Hill 15G8
Burgos 19E2
Burgundy reg. 18G3
Burhaniye 21L5
Buri 55A3
Buriram 29C6
Buritama 55A3
Buriti Alegre 55A2
Buriti Bravo 53J5
Buritirama 53J6
Buritis 55B1
Burkina Faso country 32C3
Burley 46E3
Burlington 48B1
Burlington 47I3
Burlington 47M3
Burnie 41J8
Burnley 14E5
Burns 46D3
Burnstion 14G4
Burra 41J6
Bursa 21M4
Bür Safājih 33G2
Burton upon Trent 15F6
Buru i. 29E8
Burundi country 34C4
Bururi 34C4
Buryn' 23G6
Bury St Edmunds 15H6
Busan 31C6
Bushenyi 34D4
Businga 34C3
Busto Arsizio 20C2
Buta 34C3
Butare 34C4
Butha-Buthe 37I5
Butler 48B2
Butuan 29E7
Butuinova 23I6
Buulobarde 34E3
Buxoro 26F3
Buxton 14F5
Buy 22I4
Buynaksk 23J8
Büyükmenderes r. 21L6
Buzău 21L2
Buzuluk 24G4
Byala 21K3
Byala Slatina 21J3
Byalynichy 23F5
Byaroza 11N10
Bydgoszcz 13Q4
Byerazino 23F5
Byeshankovichy 23F5
Byesville 48A3
Bykhaw 23F5
Byron Bay 42F2
Byrranga, Gory mts 25K2
Bytom 13Q5
Bytów 13P3

C

Cachoeira Alta 55A2
Caçu 55A2
Cacué 55C1
Caazapá 54E3
Cabanaconde 52D7
Cabanatuan 29E6
Cabezas 52F7
Cabimas 52D1
Cabinda 35B5
Cabinda prov. 35B5
Cabo Frio 55C3
Caboolture 42F1
Cabora Bassa, Lake resr 35D5
Caborca 46E5
Cabo San Lucas 50C4
Cabot Strait strait 45L5
Caçador 55A4
Çaçak 21I3
Caçapava do Sul 55A5
Cáceres 52F6
Cáceres 19C4
Cachoeira 55D1
Cachoeira Alta 55A2
Cao Bǎng 28C5
Capão Bonito 55A3
Capanema 53I4
Caçu 55A2
Caçué 55C1
Cáceres 33G1
Cádiz 19C5
Cádiz, Golfo de g. 19C5
Caen 15G9
Caernarfon 15C5
Caernarfon Bay 15C5
Caerphilly 15D7
Caetité 55C1
Cafelândia 55A3
Cagayan de Oro 29E7
Cagli 20C2
Cagliari 20C5
Cagliari, Golfo di g. 20C5
Cahir 17E5
Cahors 18E4
Caia 33G1
Caiapônia 55A2
Caraí 55C3
Cairngorm Mountains 16F3
Cairns 41J3
Cairo 33G1
Cajamarca 52C5
Çakovec 20G1
Calabar 32D4
Calafat 21J3
Calais 18F1
Calais 18F1
Calamocha 19F3
Calandula 35B4
Calang 29B7
Calapan 29E6
Calarasi 21L2
Calatayud 19F3
Calbayog 29E6
Calçoene 53H3
Caldas da Rainha 19B4
Caldas Novas 53I7
Caldwell 46D3
Caleta Olivia 54C7
Calexico 49E4
Calgary 44G4
Cali 52C3
California state 46C3
California, Gulf of 46E5
Callander 16E4
Callao 52C6
Callington 15C8
Caltagirone 20F6
Caltanissetta 20F6
Calvi 18I5
Camaçari 55D1
Camacupa 35B5
Camagüey 51I4
Camamu 55D1
Camana 52D7
Camapuã 55A2
Camaquã 54F4
Camaquã r. 54F4
Camarillo 49C3
Ca Mau 29C7
Ca Mau, Mui c. 29C7
Camberley 15G7
Cambodia country 29C6
Camborne 15B8
Cambrai 18F1
Cambrian Mountains hills 15D6
Cambridge 48A1
Cambridge 15H6
Cambridge 48F1
Cambridge 47J2
Cambridge 47K3
Cambridge 45H3
Cambulo 35C4
Camburgira 55B3
Camden 42D3
Camden 47I5
Cameron Park 49B1
Cameroon country 32E4
Cameroun, Mont vol. 32E4
Cametá 53I4
Camiri 52F8
Camocim 53J4
Campbell River 44F5
Campbelltown 42E5
Campbeltown 16D5
Campeche 50F5
Campeche, Bahía de g. 50F5
Camperdown 42A7
Câmpina 21K2
Campina Grande 53K5
Campinas 55B3
Campina Verde 55A2
Campo Belo 55B3
Campo Belo do Sul 55A4
Campo Grande 54F2
Campo Largo 55A4
Campo Maior 53J4
Campo Maior 19C4
Campo Mourão 54F2
Campos Altos 55B2
Campos dos Goytacazes 55C3
Campos Novos 55A4
Campos Sales 53J5
Câmpulung 21K2
Câmpulung Moldovenesc 21K1
Çan 21L4
Canada country 44H4
Çanakkale 21L4
Cananea 46E5
Canary Islands terr. 32B2
Canaveral, Cape 47K6
Canberra 42D5
Cancún 51G4
Cândido de Abreu 55A4
Canelones 54E4
Cangamba 53K5
Canguçu 54F4
Caniçado 20E6
Canindé 53K4
Çankırı 21O4
Cannes i. 16C3
Cannock 15E6
Canoas 54F3
Canoinhas 55A4
Canora 19C4
Cantábrica, Cordillera mts 19D2
Cantabrico, Mar sea 19C2
Cantagalo 55C3
Catió 55B3
Canterbury 15I7
Canterbury Bight b. 43C7
Canterbury Plains 43C6
Canton 47J4
Canton 48A2
Canto do Buriti 53J5
Canvey Island 15H7

Canyon 46G4
Cao Bǎng 28C5
Capanema 53I4
Cape Barren Island 41J8
Cape Breton Island 45L5
Cape Coast 32C4
Cape Cod Bay 48F2
Cape Girardeau 47J4
Cape Town 36D7
Cape Verde country 32[]
Cape York Peninsula 41I2
Cap-Haïtien 51J5
Capri, Isola di i. 20F4
Capricorn Channel 41K4
Caracal 21K2
Caracas 52E1
Caraguatatuba 55B3
Caraí 55C3
Carandaí 55C3
Caransebeş 21J2
Carauari 52E4
Caravaca de la Cruz 19F4
Caravelas 55D2
Carballo 48D2
Carbondale 48D2
Carbonia 20C5
Carbonita 55C2
Carcaixent 19F4
Carcassonne 18F5
Cárdenas 51H5
Cárdenas 50E4
Cardiff 15D7
Cardigan 15C6
Cardigan Bay 15C6
Cardoso 55A3
Carei 21J1
Carentan 15F9
Cariacica 55C3
Caribbean Sea 51H5
Caribou Mountains 44G4
Carinena 19F3
Carletonville 37H4
Carlisle 14E4
Carlisle 48C2
Carlos Chagas 55C2
Carlow 17F5
Carlsbad 49D4
Carlyle 16F5
Carmagnola 20B2
Carmarthen 15C7
Carmarthen Bay 15C7
Carmaux 18F4
Carmen de Patagones 54D6
Carmichael 49B1
Carmo da Cachoeira 55B3
Carmo do Paranaíba 55B2
Carnarvon 36F6
Carnarvon 40C4
Carnarvon Range hills 40E5
Carnegie, Lake salt flat 40E5
Carnot 34B3
Carnoustie 16G4
Carolina 53I5
Caroline Islands 29G7
Caroni r. 52F2
Carpathian Mountains 23C6
Carpentaria, Gulf of 41H2
Carpentras 18G4
Carpi 20D2
Carpinteria 49C3
Carrantuohill mt. 17C6
Carrara 20D2
Carrick-on-Suir 17E5
Carrick-on-Shannon 17D4
Carrizo Springs 46H6
Carroll 47I3
Carrollton 47J4
Carson City 49C1
Cartagena 51H5
Cartagena 19F5
Cartagena 52C1
Cartago 52F1
Caruaru 53K5
Carúpano 52F1
Casablanca 32C1
Casa Branca 55B3
Casa Grande 46E5
Casale Monferrato 20C2
Casas Adobes 46E5
Cascade Range mts 44F5
Cascais 19B4
Cascavel 54F2
Caserta 20F4
Casino 42F2
Casper 46F3
Caspian Lowland 23J7
Caspian Sea l. 24F5
Cassiar Mountains 44E3
Cassino 20E4
Cassongue 35B5
Castanhal 53I4
Castanhao l. 53K5
Castelli 54E5
Castello Branco 19C4
Castelnau 20E6
Castelvetrano 20E6
Castelbar 17C4
Castle Cary 15E7
Castle Douglas 16F6
Castleford 14F5
Castleisland 17C5
Castres 18F5
Castro 54B7
Castro 55A4
Castro Alves 55D1
Castro Daire 19C3
Castrovillari 20G5
Castuera 19D4
Cataguases 55C3
Catalão 55B2
Çatalca Yarımadası pen. 21M4
Cataluña aut. comm. 19G3
Catamarca 54C3
Catandica 37I1
Catanduva 55A3
Catania 20F6
Catanzaro 20G5
Catarman 29E6
Catbalogan 29E6
Catió 55B3
Cat Island 47L7
Catskill Mountains 48D1
Cattaro 55D1
Caucaia 53K4
Caucasus mts 24F5
Cauquenes 54B5
Cauta 33E1
Caviana, Ilha i. 53H3
Caxias 53J4
Caxias do Sul 55A5

Caxito 35B4
Çay 21N5
Çaycuma 21O4
Cayenne 53H3
Cayeux-sur-Mer 15I8
Çayeli 23I8
Cayman Islands terr. 51H5
Cazombo 35C5
Cebu 29E6
Cecina 20D3
Cedar City 49F2
Cedar Rapids 47I3
Ceduna 40G6
Ceerigaabo 34E2
Cegléd 21H1
Celaya 50D4
Celebes i. 29E8
Celebes Sea 38C1
Celje 13M4
Celle 13M4
Celtic Sea 12D5
Centerreach 48E2
Central, Cordillera mts 52C3
Central, Cordillera mts 52C3
Central City 46H3
Central Range mts 38E2
Central African Republic country 34B3
Central Russian Upland hills 23H5
Central Siberian Plateau 25M3
Cephalonia i. 21I5
Ceres 54D3
Ceres 55A1
Ceres 36D7
Cerignola 20F4
Çerkeş 23G8
Černvodá 21M2
Cerritos 50D4
Cerro Azul 55A4
Cerro de Pasco 52C6
Cesena 20D2
Cēsis 11N8
České Budějovice 13O6
Českomoravská vysočina hills 13O6
Český Krumlov 13O6
Çeşme 21L5
Ceuta 19D6
Cévennes mts 18F5
Chābahār 26F4
Chachapoyas 52C5
Chachoengsao 29C6
Chad country 33E3
Chadan 24K4
Chagos-charskis 21I3
Chagoda 22G4
Chaghcharān 26F3
Chai-Nat 29C6
Chajarí 54E3
Chala 52D7
Chalatenango 50G6
Chalinze 35D4
Chalkida 21J5
Châlons-en-Champagne 18G2
Chalon-sur-Saône 18G3
Chaman 26F3
Chamba 35D5
Chambersburg 48C3
Chambéry 18G4
Chamonix-Mont-Blanc 18H4
Champagnole 18G3
Champaign 47J3
Champlain, Lake 45K5
Champotón 50F5
Chamzinka 23J5
Chandeleur Islands 47J6
Chandigarh 27G3
Chandler 46E5
Chandrapur 27G5
Changbai 30B4
Changchun 30B3
Changde 27K4
Changji 31C6
Changling 30A3
Changsha 27K4
Changting 30C4
Changwon 31C6
Changzhou 28D4
Chania 21K7
Channel Islands 15E9
Channel Islands 49C4
Channel-Port-aux-Basques 45M5
Chanthaburi 29C6
Chaoyang 28D5
Chaozhou 28D5
Chapayevo 23K5
Chapayevsk 23K5
Chapecó 54F3
Chapel Hill 48C4
Chaplygin 23H5
Chapra 27H4
Charcas 50D4
Chard 15E7
Charef 19H6
Chārīkār 27F3
Charleroi 12J5
Charles City 47I3
Charleston 47L5
Charleston 47J4
Charleston 48B3
Charleville 41J5
Charleville-Mézières 18G2
Charlotte 47K4
Charlotte Amalie 51L5
Charlottesville 48B3
Charlottetown 45L5
Charlton 42A6
Charters Towers 41J4
Chase 46D1
Chashniki 23F5
Chasŏng 30B4
Château-d'Olonne 18D3
Châteaubriant 18D3
Châteaudun 18E2
Châteauroux 18E3
Château-Thierry 18F2
Châtelet 18D3
Chatham 15H7
Chatham Island 39I6
Chatham Islands 39I6
Chattanooga 47J4
Chauk 28B5
Chauny 18F2
Cheadle 15E6
Cheb 13N5
Cheboksary 22J4
Cheddar 15E7
Cheektowaga 48B1
Chegdomyn 30D2
Chegga 32C2
Chegutu 35D5
Chehalis 46C2
Chekhov 30F3
Chełm 23D6

Cheltenham 15E7
Chelva 19F4
Chelyabinsk 24H4
Chemnitz 13N5
Chengde 27K2
Chengdu 27J3
Chenzhou 27K4
Cheonan 31B5
Cheongdo 31C6
Cheongju 31B5
Chepén 52C5
Chepstow 15E7
Cher r. 18E3
Cherbourg-Octeville 15F9
Cherchell 19I5
Cherdakly 23K5
Cheremkhovo 25L4
Cheremshany 25M3
Cherepovets 22H4
Chéria 20B7
Cherkasy 23G6
Chernihiv 23G6
Chernivtsi 23E6
Chernogorsk 24K4
Chernushka 22L4
Chernyakhiv 23F6
Chernyakhovsk 11L9
Chernyanka 23H6
Chernyshevskiy 25M3
Chernyshkovskiy 23I6
Chernyye Zemli reg. 23J7
Cherry Hill 48D3
Chertkovo 23I6
Cherven Bryag 21K3
Chervonohrad 23E6
Cheryven' 23F5
Chesapeake 47L4
Chesapeake Bay 48C3
Chesham 15G7
Cheshunt 15G7
Chester 14E5
Chester 47K5
Chesterfield 14F5
Chesterfield Islands 41L3
Chester-le-Street 14F4
Chetumal 50G5
Chetwynd 44F4
Cheviot Hills 14E3
Cheyenne 46G3
Chhapra 27H4
Chhatarpur 27G4
Chhindwara 27G4
Chiang Mai 29B6
Chiang Rai 28B6
Chiayi 28D5
Chiba 31F6
Chibi 31C6
Chibougamau 45K5
Chibuto 37I2
Chicago 47J3
Chichaoua 32C1
Chichester 15G8
Chichibu 31E6
Chickasha 46H4
Chiclana de la Frontera 19C5
Chiclayo 52C5
Chico 46C4
Chicopee 48E1
Chicoutimi 45K5
Chieti 20F3
Chifeng 27K2
Chihuahua 46F6
Childers 41K5
Childress 46G5
Chile country 54B4
Chile Chico 54B7
Chililabombwe 35C5
Chilika, Ida de i. 54B6
Chilpancingo 50E5
Chiltern Hills 15G7
Chilung 28D5
Chimaltenango 50F6
Chimbas 54C3
Chimborazo mt. 52C4
Chimbote 52C5
Chimoio 35D5
Chimra 50G6
Chinandega 50G6
Chincha Alta 52C6
China country 28B4
Chinde 35D5
Chindwin r. 28B5
Chingola 35C5
Chinhoyi 35D5
Chin-ju 31C6
Chinon 18E3
Chióggia 20E2
Chipata 35D5
Chipindo 35B5
Chipman 45L5
Chippenham 15E7
Chipping Sodbury 15E7
Chirchiq 27F2
Chiredzi 35D6
Chirk 15D6
Chişinău 21M1
Chita 25M4
Chitado 35B5
Chitembo 35B5
Chitose 30F4
Chitradurga 27G5
Chitré 51H7
Chittagong 27I4
Chitungwiza 35D5
Chivasso 20B2
Chivhu 35D5
Chkalovsk 48A4
Chlef 19G5
Chobe r. 35C5
Choiseul i. 39F2
Choix 46F6
Cholet 18D3
Choluteca 51G6
Choma 35C5
Chomutov 13N5
Chon Buri 29C6
Chone 52C4
Ch'ŏngjin 30C4
Chŏngp'yŏng 31B5
Chongqing 27J4
Chonos, Archipiélago de los i. 54A6
Chorley 14E5
Chornomors'ke 21O2
Chortkiv 23E6
Chōshi 31F6
Choszczno 13O4
Choybalsan 27K2
Choyr 27J2
Christchurch 43D6
Christchurch 15F8
Christian Island terr. 39J9
Chrudim 13O6
Chudniv 23E6
Chudovo 22F4
Chuguyev 37I2
Chuhuyiv 23H6
Chukchi Sea 25T3
Chukhloma 22I4
Chukotskiy Poluostrov pen. 25T3
Chula Vista 49D4
Chulucanas 52B5
Chulym 24J4
Chumbicha 54C3

Chumphon 29B6
Chuncheon 31B5
Chungju 31B5
Chuquicamata 54C2
Chur 18I3
Churapcha 25O3
Chuxiong 27J4
Cianorte 54F2
Cide 23G8
Ciechanów 13R4
Ciego de Ávila 51I4
Ciénaga 52D1
Cienfuegos 51H4
Cieza 19F4
Cifuentes 19E3
Cilacap 29C8
Cincinnati 47K4
Çine 21M6
Cintalapa 50F5
Cirebon 29C8
Cirencester 15F7
Citluk 20G3
Citrus Heights 49B1
Città di Castello 20E3
Ciudad Acuña 50E2
Ciudad Altamirano 50D5
Ciudad Camargo 46F6
Ciudad Constitución 46E6
Ciudad del Carmen 50F5
Ciudad Delicias 46F6
Ciudad de Valles 50E4
Ciudad Guayana 52F2
Ciudad Guzmán 50D5
Ciudad Juárez 46F5
Ciudad Mante 50E4
Ciudad Obregón 46F6
Ciudad Real 19E4
Ciudad Río Bravo 46H6
Ciudad Rodrigo 19C3
Ciudad Victoria 50E4
Civitanova Marche 20E3
Civitavecchia 20D3
Çivril 21M5
Clacton-on-Sea 15I7
Clara 17E4
Claremont 48E1
Claresholm 46E1
Clarksburg 48A3
Clarksdale 47I5
Clarksville 47J4
Clearfield 48B2
Clearwater 47K6
Cleburne 47H5
Cleethorpes 14G5
Clermont 41J4
Clermont-Ferrand 18F4
Clevedon 15E7
Cleveland 48A2
Cleveland 47J4
Cleveland Heights 48A2
Cleveland Hills 14F4
Cleveleys 14D5
Clinton 47I3
Clinton 46H4
Clipperton, Île terr. 50C6
Clitheroe 14E5
Cloncurry 41I4
Clonmel 17E5
Clovis 49C2
Clovis 46G4
Cluj-Napoca 21J1
Cluses 18H3
Clwydian Range hills 14D5
Clyde r. 16F5
Clyde, Firth of est. 16E5
Clydebank 16E5
Coachella 49D4
Coalville 15F6
Coari 52E4
Coast Mountains 44F4
Coast Ranges mts 49B2
Coatbridge 16E5
Coatesville 48D3
Coatzacoalcos 50F5
Cobar 42C4
Cobden 42A7
Cobh 17D6
Cobija 52E6
Cobourg Peninsula 40G2
Cobram 42B5
Coburg 13M5
Coca 19D3
Cocalinho 55A1
Cochabamba 52E7
Cochin 55A4
Cochrane 44F5
Cocos (Keeling) Islands terr. 29B9
Codajás 52F4
Codlea 21K2
Codó 53J4
Codsall 15E6
Cody 46F3
Coeur d'Alene 46D2
Coffeyville 47H4
Coffs Harbour 42F3
Cognac 18D4
Cohoes 48E1
Coihaique 54B7
Coimbatore 27G5
Coimbra 19B3
Colac 42A7
Colatina 55C2
Colby 46G4
Colchester 15H7
Coleman 46H5
Coleraine 17F2
Colima 50D5
Coll i. 16C4
Collado Villalba 19E3
Collie 40D6
Collier Bay 40E3
Collinsville 41J4
Colmar 18H2
Colmenar Viejo 19E3
Cologne 13K5
Colombia country 52D3
Colombo 27G6
Colón 51H7
Colón 54D4
Colón 51I7
Colonial Heights 48C4
Colonsay i. 16C4
Colorado r. 46E4
Colorado state 46F4
Colorado Plateau 46E4
Colorado Springs 46G4
Colquiri 52E7
Colton 49D3
Columbia 55A1
Columbia 47J4
Columbia 47I4
Columbia 48C3
Columbia, District of admin. dist. 48C3

Columbia Mountains 44F4
Columbia Plateau 46D2
Columbus 47K5
Columbus 47J3
Columbus 47J5
Columbus 47K4
Colwyn Bay 14D5
Comacchio 20E2
Comala 54C2
Comalcalco 50F5
Comănești 21L1
Comilla 27I4
Como 20C2
Como, Lake 20C2
Comodoro Rivadavia 54C7
Comoros country 35E5
Comrat 21M1
Conakry 32B4
Conceição da Barra 55D2
Conceição do Araguaia 53I5
Conceição do Mato Dentro 55C2
Concepción 54B5
Concepción 54E2
Conchos r. 46G6
Concord 49A2
Concord 48E1
Concord 54E4
Concordia 46H4
Condeúba 55C1
Condobolin 42C4
Congleton 14E5
Congo country 34B4
Congo r. 34B4
Congo, Democratic Republic of the country 34C4
Congo Basin 34C4
Congonhas 55C3
Conneaut 48A2
Connecticut state 48E2
Connemara reg. 17C4
Conroe 47H5
Conselheiro Lafaiete 55C3
Consett 14F4
Constance, Lake 18I3
Constanța 21M2
Constantine 32D1
Conway 47I4
Coober Pedy 40G5
Cook Inlet sea chan. 44C3
Cook Islands terr. 39J3
Cook Strait strait 43E5
Cookstown 17F3
Cooktown 41J3
Coolamon 42C5
Cooma 42C5
Coonabarabran 42D3
Coonamble 42D3
Cooper Creek watercourse 41H5
Coos Bay 46C3
Copenhagen 11H9
Copertino 20H4
Copiapó 54B3
Coquimbo 54B3
Corabia 21K3
Coração de Jesus 55B2
Coraki 42F2
Coral Harbour 45J3
Coral Sea 38E3
Coral Sea Islands Territory terr. 41K3
Corby 15G6
Corcoran 49C2
Cordele 47K5
Córdoba 54D4
Córdoba 19D5
Córdoba 50E5
Córdoba, Sierras de mts 54C4
Cordova 44D3
Corfu 21H5
Corfu i. 21H5
Coria 19C4
Corigliano Calabro 20G5
Corinth 21J6
Corinth 47J5
Corinth, Gulf of sea chan. 21J5
Cork 17D6
Corleone 20E6
Çorlu 21L4
Corner Brook 45M5
Corner Inlet b. 42C7
Corning 48C1
Cornwall 45K5
Coro 52E1
Coroaci 55C2
Coroatá 53J4
Coromandel 55B2
Coromandel Coast 27H5
Coromandel Peninsula 43E3
Corona 49D4
Coronado 49D4
Coronel Fabriciano 55C2
Coronel Oviedo 54E3
Coronel Pringles 54D5
Coronel Suárez 54D5
Corowa 42C5
Corpus Christi 47H6
Corque 52E7
Corrente 53I6
Corrientes 54E3
Corrib, Lough l. 17C4
Corrientes, Cabo c. 52C2
Corrientes, Cabo c. 52C2

Côte d'Ivoire country 32C4
Cotonou 32D4
Cotopaxi, Volcán vol. 52C4
Cottbus 13O5
Cottenham 15H6
Cottian Alps mts 18H4
Council Bluffs 47I3
Courland Lagoon b. 11L9
Courtenay 44F5
Coutances 18D2
Cove Mountains hills 48B3
Coventry 15F6
Covilhã 19C3
Covington 48A4
Cowan, Lake salt flat 40E6
Cowdenbeath 16F4
Cowes 15F8
Cowra 42D4
Coxim 53H7
Cox's Bazar 27I4
Cozumel 51G4
Cradock 37G7
Craig 46F3
Craigavon 17F3
Craigieburn 42B6
Crailsheim 13M6
Craiova 21J2
Cramlington 14F3
Cranbourne 42B7
Cranbrook 44G5
Cranston 48F2
Crateús 53J5
Crato 53K5
Crawley 15G7
Credenhill 15E6
Crema 20C2
Cremona 20D2
Cres i. 20F2
Creston 46I3
Creston 47I3
Crestview 47J5
Creswick 42A6
Crete i. 21K7
Crewe 15E5
Criciúma 55A5
Crieff 16F4
Crimea disp. terr. 21O2
Cristalândia 53I6
Cristalina 55B2
Crixás 55A1
Croatia country 20G2
Crocodile r. 37I2
Cromarty Firth est. 16E3
Crook 14F4
Crookston 47H2
Crookwell 42D5
Crosby 14D5
Crotone 20G5
Crowborough 15H7
Crowland 15G6
Crows Nest 42F1
Crozet 48I3
Cruz Alta 54F3
Cruz del Eje 54D4
Cruzeiro 55B3
Cruzeiro do Sul 52D5
Crystal Brook 41H6
Crystal City 46H6
Crystal Falls 47J2
Csongrád 21I1
Cuauhtémoc 46F6
Cuba country 51I4
Cubango r. 35C5
Cubatão 55B3
Cuchilla Grande hills 54E4
Cúcuta 52D2
Cuddalore 27G5
Cuddapah 27G5
Cuenca 52C4
Cuenca 19E3
Cuernavaca 50E5
Cugir 21J2
Cuiabá 53G7
Cuillin Hills 16C3
Cuillin Sound sea chan. 16C3
Cuiluan 30C3
Culcairn 42C5
Culiacán 46F7
Cullera 19F4
Cullman 47J5
Cullybackey 17F3
Culpeper 48C3
Cumaná 52F1
Cumberland 48B3
Cumberland Plateau 47J4
Cumberland Sound sea chan. 45L3
Cumbernauld 16F5
Cumnock 16E5
Cunene r. 35B5
Cuneo 20B2
Cunnamulla 42C2
Cupar 16F4
Curaçá 53K5
Curaçao terr. 51K6
Curicó 54B4
Curitiba 55A4
Curitibanos 55A4
Currais Novos 53K5
Cururupu 53J4
Curvelo 55C2
Cusco 52D6
Cuttack 27H4
Cuxhaven 13L4
Cuyahoga Falls 48A2
Cwmbrân 15D7
Cyangugu 34C4
Cyclades is 21K6
Cyprus country 33G1
Czechia country 13N6
Częstochowa 13Q5

D

Da'an 30B3
Dabakala 32C4
Dabola 32B3
Dąbrowa Górnicza 13Q5
Dachau 13M6
Dadaab 34E3
Daegu 31C6
Daejeon 31B5
Daet 29E6
Dagana 32B3
Dagupan 29E6
Da Hinggan Ling mts 30A2
Dahlak Archipelago is 33[]
Dakar 32B3
Dākhilah, Wāḥāt ad oasis 33F2
Dakhla 32B2
Đakovo 20H2
Dalaba 32B3
Dalain Hob 27J2
Dalälven r. 11J6
Dalaman 21M6
Dalandzadgad 27J2
Đà Lạt 29C6
Dalbeattie 16F6
Dale City 48C3

Dalhart 46G4
Dali 27J4
Dalian 28E4
Dalizi 30B4
Dalkeith 16F5
Dallas 47H5
Dal'negorsk 30D3
Dal'nerechensk 30D3
Daloa 32C4
Dalton 47K5
Daly City 49A2
Daman 27G4
Damanhûr 33G1
Damascus 33G1
Damaturu 32E3
Dammam 34F1
Dampier 40D4
Dampier Archipelago is 40D4
Dampir, Selat sea chan. 29F8
Danakil reg. 33H3
Danané 32C4
Đa Năng 29C6
Danbury 48E2
Dandong 31B4
Dangriga 50G5
Danilov 22I4
Danilovka 23J6
Dankov 23H5
Danlí 51G6
Dano 32C3
Danube r. 18I2
Danube r. 23F7
Danube r. 13Q8
Danville 47J3
Danville 48C2
Danville 47L4
Daoukro 32C4
Dapaong 32D3
Dapitan 29E7
Da Qaidam 27I3
Daqing 30B3
Dara 32B3
Dar'ā 33G1
Dārāb 26E4
Darazo 32E3
Dardanelles strait 21L4
Dar es Salaam 35D4
Dargaville 43D2
Darhan 27J2
Darién, Golfo del g. 52C2
Darjiling 27H4
Darling r. 42B3
Darling Downs hills 42D1
Darling Range hills 40D6
Darlington 14F4
Darmstadt 13L6
Darnah 33F1
Daroca 19F3
Darovskoy 22J4
Dartford 15H7
Dartmoor hills 15C8
Dartmouth 45L5
Dartmouth 15D8
Daru 38E2
Darwen 14E5
Darwin 40G2
Daşkäsän 23J8
Daşoguz 26E2
Datça 21L6
Date 30F4
Datong 27K2
Daugava r. 11N8
Daugavpils 11O9
Davao 29E7
Davenport 47I3
Daventry 15F6
Daveyton 37I4
David 51H7
Davis 49B1
Davis Strait strait 45M3
Dawei 27I5
Dawmat al Jandal 26C4
Dawqah 26E5
Dawson Creek 44F4
Dax 18D5
Daylesford 42B6
Dayr az Zawr 33H1
Dayton 47K4
Daytona Beach 47K6
Dazhou 27J3
Dead Sea salt l. 33G1
Deal 15I7
Dean, Forest of 15E7
Deán Funes 54D4
Dearne r. 14F5
Death Valley depr. 49D2
Deauville 15H9
Debar 21I4
Debrecen 21I1
Debre Zeyit 34D3
Decatur 47J5
Decatur 47J4
Deccan plat. 27G5
Deception Bay 42F1
Děčín 13O5
Decorah 47I3
Dédougou 32C3
Dedovichi 22F4
Dee est. 14D5
Dee r. 15D5
Dee r. 16G3
Degema 32D4
Deggendorf 13N6
Dehloran 33H1
Dej 21J1
DeKalb 47J3
Dékoa 34B3
Delap-Uliga-Djarrit 7
Delareyville 37G4
Delaware r. 48D3
Delaware state 48D3
Delaware Bay 48D3
Delémont 18H3
Delft 12J4
Delfzijl 13K4
Delhi 27G4
Dellys 19H5
Del Mar 49D4
Delmenhorst 13L4
Delnice 20F2
De-Longa, Ostrova is 25Q2
Del Rio 46G6
Delsbo 11J6
Delta 46F4
Demba 35C4
Demidov 23F5
Deming 46F5
Demirci 21M5
Demirköy 21L4
Denbigh 14D5
Dengkou 27J2
Den Helder 12J4
Dénia 19G4
Denison 47H3
Denizli 21M6
Denman 42E4
Denmark country 11F8
Denmark Strait strait 45P3
Denny 16F4
Denpasar 29D8
Denton 47H5
D'Entrecasteaux, Point 40D6
D'Entrecasteaux Islands 41K1
Denver 46F4
Denver 48C2
Deputatskiy 25O3
Dera Ghazi Khan 27G3
Derby 15F6
Derby 48E2
Dereham 15H6
Derg, Lough l. 17D5
Dergachi 23K6
Derhachi 23H6
DeRidder 47I5
Déroute, Passage de la strait 15E9
Derry 48F1
De Rust 36F7
Derwent r. 14F6
Derwent r. 14G5
Derzhavinsk 26F1
Desē 34D2
Des Moines 47I3
Desna r. 23F6
Desnogorsk 23G5
Dessau-Roßlau 13N5
Dete 35C5
Detmold 13L5
Detroit 47K3
Detroit Lakes 47H2
Deutschlandsberg 13O7
Deva 21J2
Deventer 13K4
Devils Lake 46H2
Devizes 15F7
Devnya 21L3
Devon Island 45I2
Devonport 41J8
Devrek 21N4
Dewas 27G4
Dewsbury 14F5
Deyang 27J3
Dezfūl 33H1
Dezhneva, Mys c. 25T3
Dezhou 27K3
Dhahran 34F1
Dhaka 27I4
Dhamār 34E2
Dhanbad 27H4
Dharwad 27G5
Dhule 27G4
Dhuusa Marreeb 34E3
Diablo, Mount 49B2
Diablo Range mts 49B2
Diamante 54D4
Diamantina 55C2
Diamantina, Chapada plat. 55C1
Diamantino 53G6
Dianópolis 53I6
Diapaga 32D3
Dibaya 35C4
Dibrugarh 27I4
Dickinson 46G2
Didiéni 32C3
Diébougou 32C3
Diéma 32C3
Dieppe 15I9
Dietikon 18I3
Diffa 32E3
Digne-les-Bains 18H4
Dijon 18G3
Dikhil 34E2
Dikili 21L5
Dikson 24J2
Dila 34D3
Dili 29E8
Dillingham 44C4
Dillon 46E2
Dilolo 35C5
Dimapur 27I4
Dimbokro 32C4
Dimitrovgrad 21K3
Dimitrovgrad 23K5
Dinan 18C2
Dinant 12J5
Dinar 21N5
Dinaric Alps mts 20G2
Dindigul 27G5
Dingle Bay 17B5
Dinguiraye 32B3
Dingwall 16E3
Dioila 32C3
Dionísio Cerqueira 54F3
Diourbel 32B3
Dipayal 27H4
Dirē 32C3
Dirē Dawa 34E3
Dirk Hartog Island 40C5
Dirs 34E2
Discovery Bay 41I7
Distrito Federal admin. dist. 55B1
Ditloung 36F5
Divinópolis 55B3
Divnoye 23I7
Divo 32C4
Dixon 49B1
Dixon Entrance sea chan. 44E4
Diyarbakır 26D3
Djado, Plateau du 32E2
Djambala 34B4
Djelfa 19H6
Djenné 32C3
Djibo 32C3
Djibouti country 34E2
Djibouti 34E2
Djougou 32D4
Djoum 32E4
Dmitriyev-L'govskiy 23G5
Dmitrov 22H4
Dnieper r. 26C2
Dniester r. 23F7
Dniester r. 23F6
Dniprodzerzhyns'k 23G6
Dnipropetrovs'k 23G6
Dno 22F4
Doba 33E4
Dobele 11M8
Doberai, Jazirah pen. 29F8
Doboj 20H2
Dobrich 21L3
Dobrinka 23I5
Dobroye 23H5
Dobrush 23F5
Dodecanese is 21L7
Dodge City 46G4
Dodoma 35D4
Dogondoutchi 32D3
Doğu Menteşe Dağları mts 21M6
Doha 34F1
Dokkum 13J4
Dokshytsy 11O9
Dokuchayevs'k 23H7
Dole 18G3
Dolgellau 15D6
Dolgorukovo 23H5
Dolinsk 30F3
Dolisie 35B4
Dolomites mts 20D2
Dolores 54E5
Dolores 54E4
Dolyna 23D6
Domažlice 13N6
Dombóvár 20H1
Domeyko 54B3
Dominica country 51L5
Dominican Republic country 51J5
Domokos 21J5
Dompu 29D8
Don r. 23H7
Don r. 16G3
Donaghadee 17G3
Donald 42A6
Don Benito 19D4
Doncaster 14F5
Dondo 35B4
Donegal 17D3
Donegal Bay 17D3
Donets'k 23H7
Donets'kyy Kryazh hills 23H6
Dongchuan 27J4
Dongducheon 31B5
Dongfang 27J5
Donggang 31B5
Donghae 31C5
Đông Hơi 29C6
Dongning 30C3
Dongola 33G3
Dongting Hu l. 27K4
Dongying 27K3
Donskoye 23I7
Doomadgee 41H3
Dorbod 30B3
Dorchester 15E8
Dordogne r. 18D4
Dordrecht 12J5
Dores do Indaiá 55B2
Dori 32C3
Dorking 15G7
Dornoch Firth est. 16E3
Doro 32C3
Dorogobuzh 23G5
Dorohoi 23E7
Dorrigo 42F3
Dortmund 13K5
Dosso 32D3
Dothan 47J5
Douai 18F1
Douala 32D4
Doubtful Sound inlet 43A7
Douentza 32C3
Douglas 14C4
Douglas 46F5
Douglas 47K5
Douglas 46F3
Dourados 54F2
Douro r. 19B3
Dover 15I7
Dover 48D3
Dover 48F1
Dover 48D2
Dover 48A2
Dover, Strait of strait 15I8
Dovey r. 15D6
Downpatrick 17G3
Doylestown 48D2
Drâa, Hamada du plat. 32C2
Drachten 13K4
Drăgănești-Olt 21K2
Drăgășani 21K2
Draguignan 18H5
Drahichyn 11N10
Drakensberg mts 37I3
Drama 21K4
Drammen 11G7
Drava r. 20H2
Dréan 20B6
Dresden 13N5
Dreux 18E2
Drobeta-Turnu Severin 21J2
Drogheda 17F4
Drohobych 23D6
Droitwich Spa 15E6
Dromore 17F3
Dronfield 14F5
Drummondville 45K5
Druskininkai 11N10
Druzhnaya Gorka 11Q7
Dryanovo 21K3
Duartina 55A3
Dubai 26E4
Dubawnt Lake 45H3
Dubbo 42D4
Dublin 17F4
Dublin 47K5
Dubna 22H4
Dubno 23E6
Dubovka 23J6
Dubovskoye 23I7
Dubrovnik 20H3
Dubrovytsya 23E6
Dubuque 47I3
Dudinka 24J3
Dudley 15E6
Duékoué 32C4
Duero r. 19C3
Dufourspitze mt. 18H4
Dugi Rat 20G3
Duisburg 13K5
Dukathole 37H6
Dukhovnitskoye 23K5
Dulovo 21L3
Duluth 47I2
Dumaguete 29E7
Dumai 27J6
Dumas 46G4
Dumbarton 16E5
Dumfries 16F5
Dumyāţ 33G1
Dunajská Streda 13P7
Dunakeszi 21H1
Dunaújváros 20H1
Dunayivtsi 23E6
Duncan 46H5
Duncansby Head 16F2
Dundaga 11M8
Dundalk 17F3
Dundalk 48C3
Dundalk Bay 17F4
Dundas 48B1
Dundee 37J5
Dundee 16G4
Dundonald 17G3
Dunedin 43C7
Dunfermline 16F4
Dungannon 17F3
Dungarvan 17E5
Dungeness hd 15H8
Dungiven 17F3
Dungog 42E4
Dungu 34C3
Dungun 29C7
Dunhua 30C4
Dunkirk 18F1
Dunkirk 48B1
Dún Laoghaire 17F4
Dunmore 48D2
Dunmurry 17G3
Dunnet Head 16F2
Dunnville 48B1
Duns 16G5
Dunstable 15G7
Dupnitsa 21J3
Durağan 23G8
Durango 46G7
Durango 19E2
Durango 46F4
Durant 47H5
Durazno 54E4
Durban 37J5
Durban-Corbières 18F5
Durbanville 36D7
Durham 14F4
Durham 47L4
Durleşti 21M1
Durrës 21H4
Durrington 15F7
Dursunbey 21M5
Dushanbe 27F3
Düsseldorf 13K5
Dutse 32D3
Dutsin-Ma 32D3
Duudka, Taagga reg. 34E3
Duyun 27J4
Düzce 21N4
Dwarka 27F4
Dyat'kovo 23G5
Dyersburg 47J4
Dymytrov 23H6
Dzaoudzi 35E5
Dzerzhinsk 22I4
Dzhankoy 23G7
Działdowo 13R4
Dzuunmod 27J2
Dzyarzhynsk 11O10

E

Eagle Pass 46G6
Earn, Loch l. 16E4
Eastbourne 15H8
East China Sea 28E4
Easter Island 6
Eastern Cape prov. 37H6
Eastern Desert 33G2
Eastern Ghats mts 27G5
East Falkland i. 54E8
East Frisian Islands 13K4
East Grinstead 15G7
Easthampton 48E1
East Hartford 48E2
East Kilbride 16E5
Eastlake 48A2
Eastleigh 15F8
East Liverpool 48A2
East London 37H7
Eastmain r. 45K4
Easton 48C3
Easton 48D2
East Orange 48D2
East Providence 48F2
East Siberian Sea 25P2
East Timor country 29E8
East York 48B1
Eau Claire 47I3
Ebbw Vale 15D7
Ebebiyin 32E4
Eberswalde 13N4
Ebetsu 30F4
Eboli 20F4
Ebolowa 32E4
Ebro r. 19G3
Echizen 31E6
Écija 19D5
Eckernförde 13L3
Ecuador country 52C4
Ed 33H3
Eday i. 16G1
Ed Damazin 33G3
Ed Damer 33G3
Ed Dueim 33G3
Edéa 32E4
Edéia 55A2
Eden 42D6
Edenderry 17E4
Edessa 21J4
Edinburg 46H6
Edinburgh 16F5
Edirne 21L4
Edmonton 44G4
Edmundston 45L5
Edremit 21L5
Edward, Lake 34C4
Edwards Plateau 46G5
Effingham 47J4
Eger 13R7
Egersund 11E7
Egilsstaðir 10□2
Eğirdir 21N6
Egmont, Cape 43D4
Egvekinot 25T3
Egypt country 33G2
Ehen Hudag 27J3
Ehingen (Donau) 13L6
Eifel hills 13K5
Eigg i. 16C4
Eighty Mile Beach 40E3
Eilat 33G2
Eindhoven 12J5
Einsiedeln 18I3
Eirunepé 52E5
Eisenach 13M5
Eisenhüttenstadt 13O4
Eisenstadt 13P7
Ekenäs 11M7
Eksjö 11I8
El Aouinet 20B7
El Arrouch 20B6
Elazığ 26C3
Elba, Isola d' i. 20D3
El'ban 30E2
Elbasan 21I4
El Bayadh 32D1
Elbe r. 13L4
Elbert, Mount 46F4
Elbeuf 15I9
Elbląg 13Q3
El'brus mt. 23I8
Elburz Mountains 26D3
El Cajon 49D4
El Callao 52F2
El Campo 47H6
El Centro 49E4
El Cerro 52F7
Elche-Elx 19F4
Elda 19F4
El'dikan 25O3
Eldorado 54F3
Eldorado 55A4
El Dorado 47I5
El Dorado 47H4
Eldoret 34D3
El Ejido 19E5
Eleuthera i. 47L6
El Fasher 33F3
El Fuerte 46F6
El Geneina 33F3
Elgin 16F3
Elgin 47J3
El Goléa 32D1
Elgon, Mount 26C6
El Hadjar 20B6
El Ḥank esc. 32C2
Elhovo 21L3
Elías Piña 51J5
Elista 23J7
Elizabeth 48D2
Elizabeth City 47L4
Elizabethtown 47J4
El Jadida 32C1
El Jem 20D7
Ełk 13S4
El Kala 20C6
Elk City 46H4
El Kelaâ des Srarhna 32C1
Elkford 46E1
Elk Grove 49B1
Elkhart 47J3
Elkins 48B3
Elko 46D3
Elkton 48D3
Ellensburg 46C2
Ellesmere Island 45J2
Ellesmere Port 14E5
Ellicott City 48C3
Elliot 37H6
Ellsworth 47N3
Elmalı 21M6
El Meghaier 32D1
Elmira 48C1
Elmshorn 13L4
El Muglad 33F3
El Obeid 33G3
El Oued 32D1
El Paso 46F5
El Porvenir 46F5
El Porvenir 51I7
El Prat de Llobregat 19H3
El Progreso 50G5
El Puerto de Santa María 19C5
El Reno 46H4
El Salto 46F7
El Salvador country 50G6
El Salvador 54C3
El Tarf 20C6
El Tigre 52F2
El'ton 23J6
Elvas 19C4
Elverum 11G6
Ely 15H6
Elyria 47K3
eMalahleni 37I3
eManzimtoti 37J6
Emba 26E2
Embalenhle 37I4
Embarcación 54D2
Embu 34D4
Emden 13K4
Emet 21M5
Emgwenya 37J3
Emi Koussi mt. 33E3
Emiliano Zapata 50F5
Emirdağ 21N5
eMjindini 37J3
Emmaboda 11I8
Emmen 13K4
Emmen 18I3
eMondlo 37J4
Empangeni 37J5
Empoli 20D3
Emporia 47H4
Emporia 47L4
eMzinoni 37I4
Encarnación 54E3
Encinitas 49D4
Encruzilhada 55C1
Endeavour Strait 41I2
Endicott 48C1
Enerhodar 23G7
Engel's 23J6
England admin. div. 15E6
English Channel strait 15F9
Enid 46H4
Eniwa 30F4
Enköping 11J7
Enna 20F6
Ennis 17D5
Ennis 47H5
Enniscorthy 17F5
Enniskillen 17E3
Enschede 13K4
Ensenada 46D5
Enshi 27J3
Entebbe 34D3
Entre Rios de Minas 55B3
Entroncamento 19B4
Enugu 32D4
Envira 52D5
Ephrata 48C2
Épinal 18H2
Epsom 15G7
Equatorial Guinea country 32D4
Érd 20H1
Erdek 21L4
Erechim 54F3
Ereğli 33G1
Ereğli 21N4
Erenhot 27K2
Erfurt 13M5
'Erg Chech des. 32C2
Ērgli 11N8
Erie 48A1
Erie, Lake 48A1
Eritrea country 34D2
Erlangen 13M6
Ermelo 37I4
Ermenek 33G1
Ermoupoli 21K6
Erode 27G5
Erongo admin. reg. 36B1
Er Rachidia 32C1
Ertil' 23I6
Erzgebirge mts 13N5
Erzincan 26C3
Erzurum 26D3
Esbjerg 11F9
Escanaba 47J2
Escárcega 50F5
Eschwege 13M5
Escondido 49D4
Escuinapa 50C4
Escuintla 50F6
Eséka 32E4
Eşfahān 26E3
eSikhaleni 37K5
Eskilstuna 11J7
Eskipazar 23G8
Eskişehir 21N5
Eslāmābād-e Gharb 33H1
Eslöv 11H9
Eşme 21M5
Esmeraldas 52C3
Esperance 40E6
Esperanza 46F6
Espinhaço, Serra do mts 55C2
Espinosa 55C1
Espírito Santo state 55C2
Espírito Santo do Pinhal 55B3
Espiritu Santo i. 39G3
Espoo 11N6
Esquel 54B6
Essaouira 32C1
Essen 13K5
Essequibo r. 53G2
Essex 48C3
Es-Smara 32B2
Estância 53K6
Estcourt 37I5
Estelí 51G6
Estepona 19D5
Estevan 46G2
Estherville 47I3
Estonia country 11N7
Estrela 55A5
Estrela, Serra da mts 19C3
Estrela do Sul 55B2
Estremoz 19C4
Eswatini country 37J4
Étampes 18F2
Ethandakukhanya 37J4
Ethiopia country 34D3
Etna, Mount vol. 20F6
Etobicoke 48B1
Etosha Pan salt pan 35B5
Euclid 48A2
Euclides da Cunha 53K6
Eugene 46C3
Euphrates r. 33H1
Euphrates r. 26D3
Eura 11M6
Eureka 46C3
Eureka 46D2
Euroa 42B6
Europa, Île i. 35E6
Europa Point 19D5
Evans City 48A2
Evanston 46E3
Evansville 47J4
Evaton 37H4
Everard Range hills 40G5
Everest, Mount 27H4
Everett 46C2
Everglades swamp 47K6
Evesham 15F6
Évora 19C4
Évreux 18E2
Evvoia i. 21K5
Ewe, Loch b. 16D3
Ewo 34B4
Exe r. 15D8
Exeter 15D8
Exeter 48F1
Exmoor hills 15D7
Exmouth 40C4
Exmouth 15D8
Exmouth Gulf 40C4
Exton 48D2
Extremadura aut. comm. 19D4
Eyasi, Lake salt l. 34D4
Eyemouth 16G5
Eyjafjörður inlet 10□2
Eynsham 15F7
Eyre Peninsula 41H6
Ezakheni 37J5
Ezhva 22K3
Ezine 21L5

F

Faaborg 11G9
Fabriano 20E3
Fada-N'Gourma 32D3
Faenza 20D2
Făgăraș 21K2
Fagatogo 39I3
Fagersta 11I7
Fairbanks 44D3
Fairfax 48C3
Fairfield 49A1
Fair Head 17F2
Fair Isle i. 16H1
Fairlie 43C7
Fairmont 47I3
Fairmont 48A3
Faisalabad 27G3
Faizābād 27G3
Falenki 22K4
Falkenberg 11H8
Falkirk 16F5
Falkland Islands terr. 54E8
Falkland Sound sea chan. 54D8
Falköping 11H7
Fallbrook 49D4
Fallon 46D4
Fall River 48F2
Falmouth 15B8
Falmouth 48C3
False Bay 36D8
Falster i. 11G9
Fălticeni 23E7
Falun 11I6
Famagusta 33G1
Fandriana 35E6
Fangzheng 30C3
Fano 20E3
Faraba 32B3
Farafangana 35E6
Farāfirah, Wāḥāt al oasis 33F2
Farāh 26F3
Faranah 32B3
Fareham 15F8
Farewell, Cape 45N3
Farewell, Cape 43D5
Fargo 47H2
Faribault 47I3
Farmington 46F4
Farmville 48B4
Farnborough 15G7
Farnham 15G7
Faro 53G4
Faro 19C5
Faroe Islands terr. 10□1
Farquhar, Atoll de is 35F5
Farsund 11E7
Fasano 20G4
Fastiv 23F6
Fatehpur 27H4
Fatick 32B3
Fauske 10I3
Fawley 15F8
Faxaflói b. 10□2
Faya 33E3
Fayetteville 47I4
Fayetteville 47L4
Fdérik 32B2
Fear, Cape 47L5
Fécamp 15H9
Federalsburg 48D3
Fehmarn i. 13M3
Feijó 52D5
Feira de Santana 55D1
Feldkirch 13L7
Feldkirchen in Kärnten 13O7
Felipe C. Puerto 50G5
Felixlândia 55B2
Felixstowe 15I7
Fenoarivo Atsinanana 35E5
Feodosiya 23G7
Feres 21L4
Fergus Falls 47H2
Fériana 20C7
Ferizaj 21I3
Ferkessédougou 32C4
Fermo 20E3
Fermoselle 19C3
Fernandina Beach 47K5
Fernandópolis 55A3
Ferrara 20D2
Ferrol 19B2
Ferros 55C2
Fès 32C1
Fethiye 21M6
Fetlar i. 16□
Ffestiniog 15D6
Fianarantsoa 35E6
Fier 21H4
Fife Ness pt 16G4
Figeac 18F4
Figueira da Foz 19B3
Figueres 19H2
Figuig 32C1
Figuil 33E4
Fiji country 39H3
Filadelfia 54D2
Filey 14G4
Filingué 32D3
Filippiada 21I5
Filipstad 11I7
Fillmore 49C3
Findlay 47K3
Finger Lakes 48C1
Finike 21N6
Finisterre, Cape 19B2
Finland country 10O5
Finland, Gulf of 11M7
Finnmarksvidda reg. 10H2
Finspång 11I7
Firmat 54D4
Firminy 18G4
Fish watercourse 36C5
Fisher Strait 45J3
Fishguard 15C7
Flagstaff 37I6
Flagstaff 46E4
Flamborough Head 14G4
Flattery, Cape 46C2
Fleetwood 14D5
Flekkefjord 11E7
Flen 11J7
Flensburg 13L3
Flinders Island 41J7
Flinders Ranges mts 41H6
Flin Flon 44H4
Flint 14D5
Flint 47K3
Florence 20D3
Florence 47J5
Florence 46E5
Florence 47L5
Florencia 52C3
Flores 50G5
Flores i. 29E8
Flores, Laut sea 29D8
Floresta 53K5
Floriano 53J5
Florianópolis 55A4
Florida 54E4
Florida state 47K5
Florida, Straits of strait 47K7
Florin 49B1
Florina 21I4
Florø 11D6
Foça 21L5
Focșani 21L2
Foggia 20F4
Foix 18E5
Folda sea chan. 10I3
Foligno 20E3
Folkestone 15I7
Follonica 20D3
Fomboni 35E5
Fond du Lac 47J3
Fondi 20E4
Fonte Boa 52E4
Fontur pt 10□2
Foraker, Mount 44C3
Forchheim 13M6
Fordham 15H6
Fordingbridge 15F8
Forécariah 32B4
Forest 47J5
Forest Hill 42C5
Forestville 49A1
Forfar 16G4
Forked River 48D3
Forlì 20E2
Formby 14D5
Formiga 55B3
Formosa 54E3
Formosa 55B1
Formoso r. 55A1
Forrest City 47I4
Forssa 11M6
Forst 13O5
Fortaleza 53K4
Fort-de-France 51L6
Fort Dodge 47I3
Fort Edward 48E1
Forth r. 16F4
Forth, Firth of est. 16F4
Fort Lauderdale 47K6
Fort Macleod 46E2
Fort McMurray 44G4
Fort Myers 47K6
Fort Payne 47J5
Fort Pierce 47K6
Fort Portal 34D3
Fort Scott 47I4
Fort Smith 44G3
Fort Smith 47I4
Fort Stockton 46G5
Fort Wayne 47J3
Fort William 16D4
Fort Worth 47H5
Fossano 20B2
Foster 42C7
Fougères 18D2
Foula i. 16□
Foumban 32E4
Fouta Djallon reg. 32B3
Foveaux Strait strait 43A8
Fowler 46G4
Fox Creek 44G4
Foxe Basin g. 45K3
Foyle r. 17E3
Foyle, Lough b. 17E2
Foz do Iguaçu 54F3
Framingham 48F1
Franca 55B3
Francavilla Fontana 20G4
France country 18F3
Franceville 34B4
Francistown 35C6
Frankfort 47K4
Frankfurt (Oder) 13O4
Frankfurt am Main 13L5
Fränkische Alb hills 13M6
Franklin 48F1
Franklin 48B2
Franklin D. Roosevelt Lake resr 46D2
Frankston 42B7
Frantsa-Iosifa, Zemlya is 24G2
Frascati 20E4
Fraser r. 44F5
Fraser r. 45L4
Fraserburgh 16G3
Frauenfeld 18I3
Fray Bentos 54E4
Freckleton 14E5
Fredericia 11F9
Frederick 48C3
Fredericksburg 46H5
Fredericksburg 48C3
Fredericton 45L5
Frederikshavn 11G8
Frederiksværk 11H9
Fredonia 48B1
Fredrikstad 11G7
Freehold 48D2
Freeport 47L6
Freeport 47H6
Free State prov. 37H5
Freetown 32B4
Freiburg im Breisgau 13K6
Freising 13M6
Freistadt 13O6
Fréjus 18H5
Fremantle 40D6
Fremont 49B2
Fremont 47H3
French Guiana terr. 53H3
French Polynesia terr. 6
French Southern and Antarctic Lands terr. 7
Frenda 19G6
Fresnillo 50D4
Fresno 49C2
Freudenstadt 13L6
Fria 32B3
Frías 54C3
Fribourg 18H3
Friedrichshafen 13L7
Frobisher Bay 45L3
Frohavet b. 10F5
Frolovo 23I6
Frome 15E7
Frome, Lake salt flat 41H6
Frontera 50F5
Fronteras 46F5
Front Royal 48B3
Frosinone 20E4
Frýdek-Místek 13Q6
Fuenlabrada 19E3
Fuerte Olimpo 54E2
Fuerteventura i. 32B2
Fujairah 26E4
Fuji 31E6
Fujin 30C3
Fujinomiya 31E6
Fuji-san vol. 31E6
Fujiyoshida 31E6
Fukuchiyama 31D6
Fukui 31E5
Fukuoka 31C6
Fukushima 31F5
Fulda 13L5
Fullerton 49D4
Fulton 47I4
Fulton 48C1
Funabashi 31E6
Funafuti atoll 39H2
Funchal 32B1
Fundão 55C2
Fundão 19C3
Fundy, Bay of g. 45L5
Funtua 32D3
Furmanov 22I4
Furnas, Represa resr 55B3
Furneaux Group is 41J8
Fürstenwalde/Spree 13O4
Fürth 13M6
Fushun 30A4
Fusong 30B4
Fuyang 27K3
Fuyu 30B3
Fuyu 30B3
Fuyun 27H2
Fuzhou 28D5
Fuzhou 27K4
Fyn i. 11G9
Fyne, Loch inlet 16D5

G

Gaalkacyo 34E3
Gabela 35B5
Gabès 32E1
Gabès, Golfe de g. 32E1
Gabon country 34B4
Gaborone 37G3
Gabrovo 21K3
Gabú 32B3
Gadsden 47J5
Găești 21K2
Gaeta 20E4
Gafsa 20C7
Gagarin 23G5
Gagnoa 32C4
Gagra 23I8
Gainesville 47K6
Gainesville 47K5
Gainesville 47H5
Gainsborough 14G5
Gairdner, Lake salt flat 41H6
Galanta 13P6
Galapagos Islands 52□
Galashiels 16G5
Galați 21M2
Galatina 20H4
Galesburg 47I3
Galeshewe 36G5
Galich 22I4
Galicia aut. comm. 19C2
Galite, Canal de la sea chan. 20C6
Gallatin 47J4
Galle 27H6
Gallinas, Punta pt 52D1
Gallipoli 20H4
Gallipoli 21L4
Gällivare 10L3
Gallup 46F4
Galmudug reg. 34E3
Galtee Mountains hills 17D5
Galveston 47I6
Galveston Bay 47I6
Galway 17C4
Galway Bay 17C4
Gamalakhe 37J6
Gambēla 34D3
Gamboma 34B4
Gamleby 11J8
Gäncä 23J8
Ganda 35B5
Gandajika 35C4
Gander 45M5
Gandesa 19G3
Gandhinagar 27G4
Gandia 19F4
Ganganagar 27G4
Gangdisê Shan mts 27H3
Ganges r. 27I4
Ganges, Mouths of the 27H4
Gangneung 31C5
Gannan 30A3
Gannett Peak 44H5
Ganye 32E4
Gao 32C3
Gaoua 32C3
Gaoual 32B3
Gaoxiong 28E5
Gap 18H4
Garabogazköl Aýlagy b. 26E2
Garalo 32C3
Garanhuns 53K5
Ga-Rankuwa 37H3
Garbahaarrey 34E3
Garbsen 13L4
Garça 55A3
Garda, Lake 20D2
Garden City 46G4
Gardēz 27F3
Gardner 48F1
Gargždai 11L9
Gariep Dam resr 37G6
Garies 36C6
Garissa 34D4
Garonne r. 18D4
Garoowe 34E3
Garopaba 55A5
Garoua 32E4
Gary 47J3
Garza García 46G6
Gascony reg. 18D5
Gascony, Gulf of 18C5
Gascoyne r. 40C5
Gashua 32E3
Gaspésie, Péninsule de la pen. 47N2
Gastonia 47K4
Gatchina 11Q7
Gateshead 14F4
Gatesville 46H5
Gauteng prov. 37I4
Gavarr 23J8
Gavdos i. 21K7
Gävle 11J6
Gavrilov-Yam 22H4
Gawler 41H6
Gawler Ranges hills 41H6
Gaya 32D3
Gayéri 32D3
Gayndah 41K5
Gayny 22L3
Gaza disp. terr. 33G1
Gaza 26C3
Gaza prov. 37K2
Gaziantep 33G1
Gbarnga 32C4
Gdańsk 13Q3
Gdańsk, Gulf of 13Q3
Gdov 11O7
Gdynia 13Q3
Gedaref 33G3
Geelong 42B7
Gegen Gol 30A2
Geilo 11F6
Gejiu 27J4
Gela 20F6
Gelendzhik 23H7
Gemena 34B3
Gemlik 21M4
General Acha 54D5
General Alvear 54C5
General Juan Madariaga 54E5
General Pico 54D5
General Roca 54C5
General Salgado 55A3
General Santos 29E7
General Villegas 54D5
Geneseo 48C1
Geneva 18H3
Geneva 48C1
Geneva, Lake 18H3
Genk 13J5
Genoa 20C2
Genoa, Gulf of 20C2
Geographe Bay 40D6
Geographe Channel 40C4
Geongju 31C6
George 36F7
George Town 51H5
Georgetown 53G2
George Town 29C7
Georgetown 47L5
Georgetown 46H5
Georgia country 23I8
Georgia state 47K5
Georgian Bay 45J5
Georgiyevka 27H2
Georgiyevsk 23I7
Gera 13N5
Geral de Goiás, Serra hills 55B1
Geral do Paraná, Serra hills 55B1
Geraldton 40C5
Gerede 23G8
Germany country 13L5
Gerze 23G8
Gettysburg 48C3
Gevgelija 21J4
Geyve 21N4
Ghadāmis 32D1
Ghana country 32C4
Ghanzi 35C6
Ghanzi admin. dist. 36F2
Ghardaïa 32D1
Gharyān 33E1
Ghazal, Bahr el watercourse 33E3
Ghazaouet 19F6
Ghaziabad 27G4
Ghaznī 26F3
Ghent 12I5
Gheorgheni 21K1
Gherla 21J1
Ghisonaccia 18I5
Giaginskaya 23I7
Giannitsa 21J4
Giant's Causeway lava field 17F2
Giarre 20F6
Gibraltar terr. 19D5
Gibraltar, Strait of strait 19C6
Gibson Desert 40E4
Gießen 13L5
Gifu 31E6
Gijón/Xixón 19D2
Gila r. 49E4
Gilbués 53I5
Gilgandra 42D3
Gilgit 27G3
Gillette 46F3
Gillingham 15H7
Gilroy 49B2
Gimcheon 31C5
Gimhae 31C6
Ginosa 20G4
Gioia del Colle 20G4
Gippsland reg. 42B7
Giresun 23H8
Girona 19H3
Gironde est. 18D4
Girvan 16E5
Gisborne 43G4
Gislaved 11H8
Gitarama 34C4
Gitega 34C4
Giulianova 20E3
Giurgiu 21K3
Giyani 37J2
Giza 33G2
Gjakovë 21I3
Gjilan 21I3
Gjirokastër 21I4
Gjøvik 11G6
Glace Bay 45M5
Gladstone 41K4
Glamoč 20G2
Glasgow 16E5
Glasgow 47J4
Glastonbury 15E7
Glazov 22L4
Glendale 46E5
Glendale 49C3
Glen Innes 42E2
Glen More valley 16E3
Glenrothes 16F4
Glens Falls 48E1
Gliwice 13Q5
Globe 46E5
Głogów 13P5
Glomfjord 10H3
Gloucester 42E3
Gloucester 15E7
Gloucester 48F1
Gloversville 48D1
Glubokiy 23I6
Glubokoye 27H1
Gmünd 13O6
Gmunden 13N7
Gniezno 13P4
Goba 34E3
Gobabis 36D2
Gobi Desert 27J2
Gobō 31D6
Gochas 36D3
Godalming 15G7
Godavari r. 27H5
Goiana 53L5
Goianésia 55A1
Goiânia 55A2
Goiás 55A1
Goiás state 55A2
Goioerê 54F2
Gökçeada i. 21K4
Gölcük 21M4
Gold Coast 42F2
Golden Bay 43D5
Goldsboro 47L4
Göle 23I8
Goleta 49C3
Gölköy 23H8
Golmud 27I3
Gölpazarı 21N4
Goma 34C4
Gombe 32E3
Gómez Palacio 46G6
Gonaïves 51J5
Gonbad-e Kāvūs 26E3
Gonder 34D2
Gönen 21L4
Gonzales 47H6
Good Hope, Cape of 36D8
Goole 14G5
Goondiwindi 42E2
Göppingen 13L6
Gorakhpur 27H4
Gördes 21M5
Gorē 34D3
Gorey 17F5
Gorgān 26E3
Gorizia 20E2
Gorlice 23D6
Görlitz 13O5
Gorna Oryahovitsa 21K3
Gornji Milanovac 21I2
Gornji Vakuf 20G3
Gorno-Altaysk 24J4
Gornozavodsk 30F3
Gornyak 24J4
Gornyy 23K6
Gorodets 22I4
Gorodishche 23J5
Gorodishche 23J6
Gorodovikovsk 23I7
Goroka 38E2
Gorokhovets 22I4
Gorom Gorom 32C3
Gorontalo 29E7
Gorshechnoye 23H6
Goryachiy Klyuch 23H7
Goryeong 31C6
Gorzów Wielkopolski 13O4
Gosforth 14F3
Goshen 48D2
Gospić 20F2
Gosport 15F8
Gossi 32C3
Gostivar 21I4
Götene 11H7
Gotha 13M5
Gothenburg 11G8
Gotland i. 11K8
Gotse Delchev 21J4
Gōtsu 31D6
Göttingen 13L5
Gouda 12J4
Gouin, Réservoir resr 45K5
Goulburn 42D5
Goundam 32C3
Gouraya 19G5
Gourcy 32C3
Gouré 32E3
Governador Valadares 55C2
Govĭ Altayn Nuruu mts 27I2
Goya 54E3
Göyçay 23J8
Gozo i. 20F6
Graaff-Reinet 36G7
Grabouw 36D8
Gračac 20F2
Grafton 42F2
Grafton 48A3
Graham 46H5
Grahamstown 37H7
Grajaú 53I5
Grampian Mountains 16E4
Granada 51G6
Granada 19E5
Granby 45K5
Gran Canaria i. 32B2
Gran Chaco reg. 54D3
Grand Bahama i. 47L6
Grand Bank 45M5
Grand Banks of Newfoundland sea feature 45M5
Grand-Bassam 32C4
Grand Canyon gorge 49F2
Grand Cayman i. 51H5
Grande, Bahía b. 54C8
Grande Prairie 44G4
Grand Erg Occidental des. 32C1
Grand Erg Oriental des. 32D2
Grandes, Salinas salt marsh 54C4
Grand Falls-Windsor 45M5
Grand Forks 47H2
Grand Island 46H3
Grand Junction 46F4
Grand-Lahou 32C4
Grândola 19B4
Grand Rapids 47J3
Grand Rapids 47I2
Grand Turk 51J4
Grängesberg 11I6
Granja 53J4
Gränna 11I7
Grantham 15G6
Grantown-on-Spey 16F3
Grants 46F4
Grants Pass 46C3
Grão Mogol 55C2
Graskop 37J3
Grasse 18H5
Graus 19G2
Gravataí 55A5
Gravesend 15H7
Gravina in Puglia 20G4
Grays 15H7
Graz 13O7
Great Abaco i. 47L6
Great Australian Bight g. 40F6
Great Bahama Bank sea feature 47L6
Great Barrier Island 43E3
Great Barrier Reef reef 41I2
Great Basin 46D4
Great Bear Lake 44G3
Great Belt sea chan. 11G9
Great Bend 46H4
Great Britain i. 12G4
Great Dividing Range mts 42B6
Greater Antilles is 51H4
Great Exuma i. 47L7
Great Falls 46E2
Great Inagua i. 47M7
Great Karoo plat. 36F7
Great Malvern 15E6
Great Nicobar i. 27I6
Great Ouse r. 15H6
Great Rift Valley valley 34D4
Great Salt Lake 46E3
Great Salt Lake Desert 46E3
Great Sand Sea des. 33F2
Great Sandy Desert 40E4
Great Slave Lake 44G3
Great Stour r. 15I7
Great Torrington 15C8
Great Victoria Desert 40F5
Great Waltham 15H7
Great Yarmouth 15I6
Greece country 21I5
Greece 48C1
Greeley 46G3
Green Bay 47J3
Green Bay 47J2
Greeneville 47K4
Greenfield 48E1
Greenland terr. 45N2
Greenland Sea 24A2
Greenock 16E5
Green River 46F3
Greensburg 48B2
Greenville 32C4
Greenville 47J5
Greenville 47I5
Greenville 47K5
Greenville 47H5
Greenwich 48E2
Greenwood 47K5
Gregory, Lake salt flat 40F4
Gregory Range hills 41I3
Gregory Range hills 40E4
Greifswald 13N3
Grenaa 11G8
Grenada 47J5
Grenada country 51L6
Grenfell 42D4
Grenoble 18G4
Gretna 16F6
Grevena 21I4
Greybull 46F3
Greymouth 43C6
Grey Range hills 42A2
Gribanovskiy 23I6
Griffin 47K5
Griffith 42C5
Grimari 34C3
Grimsby 14G5
Grimshaw 44G4
Grimstad 11F7
Grindavík 10□2
Grindsted 11F9
Grmeč mts 20G2
Groblersdal 37I3
Groningen 13K4
Groote Eylandt i. 41H2
Grootfontein 35B5
Groot Swartberge mts 36E7
Grootvloer salt pan 36E5
Grosseto 20D3
Groß-Gerau 13L6
Großglockner mt. 13N7
Grover Beach 49B3
Groznyy 23J8
Grubišno Polje 20G2
Grudziądz 13Q4
Gryazi 23H5
Gryazovets 22I4
Gryfice 13O4
Gryfino 13O4
Guadalajara 50D4
Guadalajara 19E3
Guadalcanal i. 41M1
Guadalquivir r. 19C5
Guadalupe Victoria 46G7
Guadarrama, Sierra de mts 19D3
Guadeloupe terr. 51L5
Guadix 19E5
Guaíba 55A5
Guaíra 54F2
Gualeguay 54E4
Gualeguaychu 54E4
Guam terr. 29G6
Guamúchil 46F6
Guanajuato 50D4
Guanambi 55C1
Guanare 52E2
Guane 51H4
Guangyuan 27J3
Guangzhou 27K4
Guanhães 55C2
Guantánamo 51I4
Guapé 55B3
Guaporé r. 52E6
Guaporé 55A5
Guarabira 53K5
Guaranda 52C4
Guarapari 55C3
Guarapuava 55A4
Guararapes 55A3
Guaratinguetá 55B3
Guaratuba 55A4
Guarda 19C3
Guarujá 55B3
Guasave 46F6
Guatemala country 50F5
Guatemala City 50F6
Guaxupé 55B3
Guayaquil 52C4
Guayaquil, Golfo de g. 52B4
Guaymas 46E6
Gubkin 23H6
Gudermes 23J8
Guéckédou 32B4
Guelma 20B6
Guelmim 32B2
Guelph 48A1
Guéret 18E3
Guernsey terr. 15E9
Guérou 32B3
Guider 33E4
Guidonia Montecelio 20E4
Guiglo 32C4
Guildford 15G7
Guilin 27K4
Guimarães 53J4
Guimarães 19B3
Guinea country 32B3
Guinea, Gulf of 32D4
Guinea-Bissau country 32B3
Güines 51H4
Guiratinga 53H7
Guiyang 27J4
Gujranwala 27G3
Gukovo 23H6
Gulbene 11O8
Gulfport 47J5
Guliston 27F2
Gul'kevichi 23I7
Gulu 34D3
Gumare 35C5
Gumel 32D3
Gumi 31C5
Guna 27G4
Gundagai 42D5
Güney 21M5

Gunnison 46F4
Gunsan 31B6
Guntakal 27G5
Gunungsitoli 29B7
Gurinhatã 55A2
Gurupá 53H4
Gurupi 53I6
Gur'yevsk 11L9
Gusau 32D3
Gusev 11M9
Gushan 31A5
Gusinoozersk 25L4
Gus'-Khrustal'nyy 22I5
Güstrow 13N4
Gütersloh 13L5
Guwahati 27I4
Guyana *country* 53G2
Guymon 46G4
Guyra 42E3
G'uzor 26F3
Gvardeysk 11L9
Gwalior 27G4
Gwanda 35C6
Gwangju 31B6
Gwardafuy, Gees *c.* 34F2
Gweru 35C5
Gweta 35C6
Gwoza 32E3
Gydan Peninsula 24I2
Gympie 41K5
Gyöngyös 13Q7
Gyr 20G1
Gytheio 21J6
Gyula 21I1
Gyumri 23I8

H

Haapsalu 11M7
Haarlem 12J4
Ḥabbān 34E2
Hachinohe 30F4
Hackensack 48D2
Haddington 16G5
Hadejia 32E3
Haderslev 11F9
Hadyach 23G6
Haeju 31B5
Haenam 31B6
Hagåtña 29G6
Hagen 13K5
Hagerstown 48C3
Hagfors 11H6
Hagi 31C6
Hague, Cap de la *c.* 15F9
Haguenau 18H2
Haida Gwaii 44E4
Haifa 33G1
Haikou 27K4
Ḥā'il 26D4
Hailin 30C3
Hailsham 15H8
Hailun 30B3
Hainan Dao *i.* 27K5
Hai Phong 28C5
Haiti *country* 51J5
Hajdúböszörmény 21I1
Ḥajjah 34E2
Hakodate 30F4
Hakui 31E5
Halaib Triangle *terr.* 33G2
Halberstadt 13M5
Halden 11G7
Haldensleben 13M4
Halesowen 15E6
Halesworth 15I6
Halifax 45L5
Halifax 14F5
Halle (Saale) 13M5
Hällefors 11I7
Hallein 13N7
Hallsberg 11I7
Halls Creek 40F3
Halmahera *i.* 29E7
Halmstad 11H8
Hals 11G8
Haltwhistle 14E4
Hamada 31D6
Hamadān 33H1
Ḥamāh 33G1
Hamamatsu 31E6
Hamar 11G6
Hambantota 27H6
Hamburg 13L4
Hamburg 48B1
Hamden 48E2
Hämeenlinna 11N6
Hameln 13L4
Hamersley Range *mts* 40D4
Hamhŭng 31B5
Hami 27I2
Hamilton 51L2
Hamilton 48B1
Hamilton 43E3
Hamilton 16E5
Hamilton 47K4
Hamina 11O6
Hamju 31B5
Hamm 13K5
Hammamet 20D6
Hammamet, Golfe de *g.* 20D6
Hammerfest 10M1
Hammonton 48D3
Hampshire Downs *hills* 15F7
Hampton 48C4
Ḥanak 34D1
Hanamaki 31F5
Handan 27K3
Handeni 35D4
Hanford 49C2
Hangayn Nuruu *mts* 27I2
Hangzhou 28E4
Hangzhou Wan *b.* 28E4
Hanko 11M7
Hanna 44G4
Hannibal 47I4
Hannover 13L4
Hanöbukten *b.* 11I9
Ha Nôi 28C5
Hanover 48C3
Hantsavichy 11O10
Hanzhong 27J3
Haparanda 10N4
Happy Valley-Goose Bay 45L4
Ḥaraḍh 34E1
Haradok 23F5
Haramachi 31F5
Harare 35D5
Harbin 30B3
Hardangerfjorden *sea chan.* 11D7
Hardap *admin. reg.* 36C3
Härer 34E3
Hargeysa 34E3
Harima-nada *b.* 31D6
Harjavalta 11M6
Harleston 15I6
Harlow 15H7
Harmanli 21K4
Harney Basin 46C3
Härnösand 10J5
Harper 32C4
Harris, Sound of *sea chan.* 16B3
Harrisburg 48C2
Harrison 47I4
Harrisonburg 48B3
Harrisonville 47I4
Harrogate 14F5
Hârșova 21L2
Harstad 10J2
Hartberg 13O7
Hartford 48E2
Hartlepool 14F4
Harvey 46G2
Harwich 15I7
Haskovo 21K4
Haslemere 15G7
Hassan 27G5
Hasselt 12J5
Hässleholm 11H8
Hastings 43F4
Hastings 15H8
Hastings 47I3
Hastings 46H3
Hatfield 14G5
Hatteras, Cape 47L4
Hattiesburg 47J5
Hat Yai 29C7
Haud *reg.* 34E3
Haugesund 11D7
Haukivesi *l.* 10P5
Hauraki Gulf 43E3
Haut Atlas *mts* 32C1
Haute-Normandie *admin. reg.* 15I9
Hauts Plateaux 32C1
Havana 51H4
Havant 15G8
Haverfordwest 15C7
Haverhill 48F1
Havlíčkův Brod 13O6
Havran 21L5
Havre 46F2
Havre Rock *i.* 39I5
Havza 23G8
Hawai'i *i.* 46□
Hawick 16G5
Hawke Bay 43F4
Haxby 14F4
Hay 42B5
Hay *watercourse* 41H4
Haymā' 26E5
Hayrabolu 21L4
Hay River 44G3
Hays 46H4
Haysyn 23F6
Hayward 49A2
Haywards Heath 15G8
Hazleton 48D2
Ḥazm al Jawf 34E2
Heanor 15F5
Heard Island and McDonald Islands *terr.* 7
Hearst 45J5
Heathcote 42B6
Hechi 27J4
Hedemora 11I6
Hefei 27K3
Hegang 30C3
Heide 13L3
Heidelberg 13L6
Heidelberg 37I4
Heihe 30B2
Heilbronn 13L6
Heilongjiang *prov.* 30C3
Heinola 11O6
Hekla *vol.* 10□
Helena 46E2
Helensburgh 16E4
Helgoländer Bucht *g.* 13L3
Hellín 19F4
Helmand *r.* 26F3
Helmond 13J5
Helmsdale 16F2
Helmstedt 13M4
Helong 30C4
Helsingborg 11H8
Helsingør 11H8
Helsinki 11N6
Hemel Hempstead 15G7
Hemet 49D4
Hendek 21N4
Henderson 47L4
Henderson 49E2
Henderson 47I5
Hengelo 13K4
Hengshan 30C3
Hengyang 27K4
Heniches'k 23G7
Henley-on-Thames 15G7
Herāt 26F3
Hereford 15E6
Hereford 46G5
Herford 13L4
Herisau 18I3
Herkimer 48D1
Hermes, Cape 37I6
Hermosillo 46E6
Hernandarias 54F3
Herne Bay 15I7
Herning 11F8
Hérouville-St-Clair 15G9
Herrera del Duque 19D4
Hershey 48C2
Hertford 15G7
Hespería 49D3
Hettstedt 13M5
Hexham 14E4
Heysham 14E4
Heywood 14E5
Heze 27K3
Hibbing 47I2
Hidalgo del Parral 46F6
Hidrolândia 55A2
Highlands 48E2
Highland Springs 48C4
High Level 44G4
High Point 47L4
High Prairie 44G4
High Wycombe 15G7
Higüey 51K5
Hiiumaa *i.* 11M7
Hijaz *reg.* 34D1
Hikone 31E6
Hildesheim 13L4
Hillah 33H1
Hillerød 11H9
Hillston 42B4
Hilo 46□
Hilton Head Island 47K5
Hilversum 12J4
Himalaya *mts* 27G3
Himarë 21H4
Himeji 31D6
Hinckley 15F6
Hindley 14E5
Hindu Kush *mts* 26F3
Hinesville 47K5
Hinthada 27I5
Hirosaki 30F4
Hiroshima 31D6
Hirson 18G2
Hirtshals 11F8
Hisar 27G4
Hispaniola *i.* 51J4
Hitachi 31F5
Hitachinaka 31F5
Hjälmaren *l.* 11I7
Hjo 11I7
Hjørring 11G8
Hlotse 37I5
Hlukhiv 23G6
Hlybokaye 11O9
Ho 32D4
Hobart 41J8
Hobbs 46G5
Hobro 11F8
Ho Chi Minh City 29C6
Hoddesdon 15G7
Hodeidah 34E2
Hódmezővásárhely 21I1
Hoeryŏng 30C4
Hof 13M5
Hofors 11J6
Hofsjökull *ice cap* 10□
Hōfu 31C6
Höganäs 11H8
Hohhot 27K2
Hoh Xil Shan *mts* 27H3
Hoima 34D3
Hokkaidō *i.* 30F4
Hokksund 11F7
Holbæk 11G9
Holbrook 42C5
Holdrege 46H3
Holguín 51I4
Hollidaysburg 48B2
Hollister 49B2
Holly Springs 47J5
Hollywood 49C4
Hollywood 47K6
Holmestrand 11G7
Holmsund 10L5
Holstebro 11F8
Holt 15I6
Holyhead 14C5
Holyhead Bay 14C5
Holy Island 14F3
Holy Island 14C5
Hombori 32C3
Home Bay 45L3
Homer 48C1
Homs 33G1
Homyel' 23F5
Honaz 21M6
Hondo 46H6
Honduras *country* 51G6
Hønefoss 11G6
Hongjiang 27J4
Hong Kong 27K4
Hongwŏn 31B4
Hongze Hu *l.* 28D4
Honiara 41L1
Honiton 15D8
Honolulu 46□
Honshū *i.* 31D6
Hoogeveen 13K4
Hoopstad 37G4
Höör 11H9
Hoorn 12J4
Hoover Dam 49E2
Hopa 23I8
Hope 47I5
Hopewell 48C4
Hopkinsville 47J4
Hörby 11H9
Horki 23F5
Horlivka 23H6
Hormuz, Strait of *strait* 26E4
Horn 13O6
Horn, Cape 54C9
Horncastle 14G5
Hörnefors 10K5
Hornsby 42E4
Horodenka 23E6
Horodnya 23F6
Horodok 23E6
Horodok 23D6
Horsens 11F9
Horsham 41I7
Horsham 15G7
Horten 11G7
Hoshiarpur 27G3
Hotan 27H3
Hot Springs 47I5
Houghton 47J2
Houghton le Spring 14F4
Houma 47I6
Houston 47H6
Hovd 27I2
Hove 15G8
Hovmantorp 11I8
Hövsgöl Nuur *l.* 27J1
Howden 14G5
Howe, Cape 42D6
Howick 37J5
Howland Island *terr.* 39I1
Howlong 42C5
Höxter 13L5
Hoy *i.* 16F2
Hoyerswerda 13O5
Hradec Králové 13O5
Hrazdan 23I8
Hrebinka 23G6
Hrodna 11M10
Huacho 52C6
Huadian 30B4
Huaibei 27K3
Huaihua 27J4
Huainan 27K3
Huajuápan de León 50E5
Huambo 35B5
Huanan 30C3
Huancavelica 52C6
Huancayo 52C6
Huanren 30B4
Huánuco 52C5
Huaraz 52C5
Huarmey 52C6
Huasco 54B3
Huatabampo 46F6
Hubballi 27G5
Hucknall 15F5
Huddersfield 14F5
Huder 30A2
Hudiksvall 11J6
Hudson 48F1
Hudson 48E1
Hudson *r.* 48E2
Hudson Bay *sea* 45J4
Hudson Falls 48E1
Hudson Strait *strait* 45K3
Huê 29C6
Huehuetenango 50F5
Huelva 19C5
Huesca 19F2
Hughenden 41I4
Hughson 49B2
Hugo 47H5
Huhudi 36G4
Huib-Hoch Plateau 36C4
Huíla, Planalto da 35B5
Huimanguillo 50F5
Huittinen 11M6
Hulan 30B3
Hulan Ergi 30A3
Ḥulayfah 34E1
Hulin 30D3
Hull 45K5
Hultsfred 11I8
Hulun Buir 27K2
Hulwān 33G2
Humaitá 52F5
Humber, Mouth of the 14H5
Hume Reservoir 42C5
Hün 33E2
Húnaflói *b.* 10□
Hunedoara 21J2
Hungary *country* 20H1
Hŭngnam 31B5
Hunstanton 15H6
Huntingdon 15G6
Huntingdon 48C2
Huntington 47J3
Huntington 47K4
Huntington Beach 49C4
Huntly 16G3
Huntsville 45K5
Huntsville 47J5
Huntsville 47H5
Hurghada 33G2
Huron 46H3
Huron, Lake 48A1
Húsavík 10□
Huşi 21M1
Husnes 11D7
Husum 13L3
Husum 10K5
Hutchinson 46H4
Huzhou 28E4
Hvar *i.* 20G3
Hwange 35C5
Hyderabad 27G5
Hyderabad 26F4
Hyères 18H5
Hyesan 30C4
Hythe 15I7
Hyūga 31C6
Hyvinkää 11N6

I

Iaçu 55C1
Ianca 21L2
Iași 21L1
Ibadan 32D4
Ibagué 52C3
Ibaiti 55A3
Ibarra 52C3
Ibb 34E2
Iberian Peninsula 19
Ibiá 55B2
Ibiapaba, Serra da *hills* 53J4
Ibiassucê 55C1
Ibicaraí 55D1
Ibirama 55A4
Ibitinga 55A3
Ibiza 19G4
Ibiza *i.* 19G4
Ibotirama 53J6
Ibrā' 26E4
Ibrī 26E4
Ica 52C6
Icatu 53J4
Iceland *country* 10□
Ichinomiya 31E6
Ichinoseki 31F5
Ichnya 23G6
Icó 53K5
Iconha 55C3
Idah 32D4
Idaho *state* 46E3
Idaho Falls 46E3
Idar-Oberstein 13K6
Idfū 33G2
Idlib 33G1
Iepê 55A3
Ifakara 35D4
Ifanadiana 35E6
Ife 32D4
Ifenat 33E3
Iférouâne 32D3
Ifôghas, Adrar des *hills* 32D3
Iganga 33G4
Igarapava 55B3
Igarka 24J3
Iggesund 11J6
Iglesias 20C5
Ignalina 11O9
Igoumenitsa 21I5
Igra 22L4
Igrim 22M3
Iguaçu *r.* 55A4
Iguaí 55C1
Iguala 50E5
Igualada 19G3
Iguape 55B4
Iguatemi 54F2
Iguatu 53K5
Ihosy 35E6
Ihtiman 21J3
Ih Tol Gol 30A2
Iisalmi 10O5
Iizuka 31C6
Ijebu-Ode 32D4
Ijevan 23I8
IJssel *r.* 13J4
Ikageng 37H4
Ikare 32D4
Ikaria *i.* 21L6
Ikast 11F8
Ikom 32D4
Iksan 31B6
Ikungu 35D4
Ilagan 29E6
Īlām 33H1
Iława 13Q4
Ilebo 35C4
Ilemi Triangle *terr.* 34D3
Ilford 15H7
Ilfracombe 15C7
Ilgaz 23G8
Ilha Grande *resr* 54F2
Ilhavo 19B3
Ilhéus 55D1
Ilkeston 15F6
Ilkley 14F5
Illapel 54B4
Illéla 32D3
Iller *r.* 13L6
Illichivs'k 21N1
Illinois *r.* 47I4
Illinois *state* 47J4
Illizi 32D2
Il'men', Ozero *l.* 22F4
Ilminster 15E8
Ilo 52D7
Iloilo 29E6
Ilorin 32D4
Ilovlya 23I6
Ilulissat 45M3
Imabari 31D6
Imari 31C6
Imaruí 55A5
Imatra 11P6
Imbituva 55A4
Imola 20D2
Imperatriz 53I5
Imperia 20C3
Imperial 49E4
Imperial Beach 49D4
Impfondo 34B3
Imphal 27I4
İmroz 21K4
Ina 31E6
Inarijärvi *l.* 10O2
Inca 19H4
Incheon 31B5
Indaal, Loch *b.* 16C5
Indalsälven *r.* 10J5
Inda Silasē 34D2
Independence 47I4
Inderbor 26E2
India *country* 27G4
Indiana 48B2
Indiana *state* 47J3
Indianapolis 47J4
Indianola 47I3
Indianola 47I5
Indigirka *r.* 25P2
Indija 21I2
Indio 49D4
Indonesia *country* 29D8
Indore 27G4
Indus *r.* 27F4
Indus, Mouths of the 26F4
İnebolu 23G8
İnegöl 21M4
Inglewood 42E2
Inglewood 49C4
Ingoldmells 14H5
Ingolstadt 13M6
Inhambane 37L2
Inhambane *prov.* 37L2
Inhumas 55A2
Inírida 52E3
Inkeroinen 11O6
Inner Sound *sea chan.* 16D3
Innisfail 41J3
Innsbruck 13M7
Inongo 34B4
Inowrocław 13Q4
In Salah 32D2
Inscription, Cape 40C5
Inta 24H3
International Falls 47I2
Inuvik 44E3
Invercargill 43B8
Invergordon 16E3
Inverkeithing 16F4
Inverness 16E3
Investigator Group *is* 40G6
Investigator Strait 41H7
Inyonga 35D4
Inza 23J5
Inzhavino 23I5
Ioannina 21I5
Iola 47H4
Iona *i.* 16C4
Ionian Islands 21H5
Ionian Sea 20H5
Ios *i.* 21K6
Iowa *state* 47I3
Iowa City 47I3
Ipameri 55A2
Ipanema 55C2
Ipatinga 55C2
Ipatovo 23I7
Ipelegeng 37G4
Ipiales 52C3
Ipiaú 55D1
Ipirá 55D1
Ipiranga 55A4
Ipoh 29C7
Iporá 55A2
Ippy 34C3
Ipsala 21L4
Ipswich 42F1
Ipswich 15I6
Ipu 53J4
Iqaluit 45L3
Iquique 54B2
Iquitos 52D4
Īrāf 54F3
Iraklion 21K7
Iramaia 55C1
Iran *country* 26E3
Īrānshahr 26F4
Irapuato 50D4
Iraq *country* 33H1
Irará 55D1
Irati 55A4
Irbid 33G1
Irbit 24H4
Irecê 53J6
Ireland *country* 17E4
Ireland *i.* 17
Iringa 35D4
Iriri *r.* 53H4
Irish Sea 17G4
Irituia 53I4
Irkutsk 25L4
Irondequoit 48C1
Iron Mountain 47J2
Irosin 29E6
Irpin' 23F6
Irrawaddy *r.* 27I5
Irtysh *r.* 27G1
Irun 19F2
Irvine 16E5
Irvine 49D4
Isa 32D3
Isabela 29E7
Ísafjarðardjúp *est.* 10□
Ísafjörður 10□
Ise 31E6
Isère *r.* 18G4
Isernia 20F4
Ise-wan *b.* 31E6
Iseyin 32D4
Ishinomaki 31F5
Ishioka 31F5
Isil'kul' 24I4
Isipingo 37J5
Isiro 34C3
Iskenderun 33G1
Iskitim 24J4
Islamabad 27G3
Islands, Bay of 43E2
Islay *i.* 16C5
Isoka 35D5
Isparta 21N6
Isperih 21L3
Israel *country* 33G1
Issia 32C4
Issoire 18F4
İstanbul 21M4
Istiaia 21J5
Istres 18G5
Istria *pen.* 20E2
Itaberá 55A3
Itaberaba 55C1
Itaberaí 55A2
Itabira 55C2
Itabirito 55C3
Itabuna 55D1
Itacajá 53I5
Itacarambi 55B1
Itacoatiara 53G4
Itaeté 55C1
Itaguaçu 55C2
Itaí 55A3
Itaiópolis 55A4
Itaituba 53G4
Itajaí 55A4
Itajubá 55B3
Itajuípe 55D1
Italy *country* 20E3
Itamarandiba 55C2
Itambé 55C1
Itanhaém 55B4
Itanhém 55C2
Itaobim 55C2
Itapaci 55A1
Itapajipe 55A2
Itapebi 55D1
Itapecerica 55B3
Itapecuru Mirim 53J4
Itapemirim 55C3
Itaperuna 55C3
Itapetinga 55C1
Itapetininga 55A3
Itapeva 55A3
Itapicuru 53J6
Itapipoca 53K4
Itapira 55B3
Itaporanga 55A3
Itapuã 55A5
Itaqui 54E3
Itararé 55A4
Itatiba 55B3
Itaúna 55B3
Itea 21J5
Ithaca 48C1
Itinga 55C2
Itiquira 53H7
Itiruçu 55C1
Itō 31E6
Itu 55B3
Ituaçu 55C1
Ituberá 55D1
Ituiutaba 55A2
Itumbiara 55A2
Itupiranga 53I5
Ituporanga 55A4
Iturama 55A2
Itzehoe 13L4
Iul'tin 25T3
Ivalo 10O2
Ivanava 11N10
Ivankiv 23F6
Ivano-Frankivs'k 23E6
Ivanovka 30B2
Ivanovo 22I4
Ivanteyevka 23K5
Ivatsevichy 11N10
Ivaylovgrad 21L4
Ivdel' 24H3
Ivrea 20B2
İvrindi 21L5
Ivujivik 45K3
Ivyanyets 11O10
Iwaki 31F5
Iwakuni 31D6
Iwamizawa 30F4
Iwo 32D4
Iwye 11N10
Ixmiquilpán 50E4
Ixtlán 50D4
Ixworth 15H6
Izberbash 23J8
Izhevsk 24G4
İzmayil 21M2
İzmir 21L5
İzmir Körfezi *g.* 21L5
İzmit 21M4
Izozog, Bañados del *swamp* 52F7
Iztochni Rodopi *mts* 21K4
Izumo 31D6
Izyaslav 23E6
Iz"yayu 22M2
Izyum 23H6

J

Jabalpur 27G4
Jablanica 20G3
Jaboatão dos Guararapes 53L5
Jaboticabal 55A3
Jacaraci 55C1
Jacareí 55B3
Jacarezinho 55A3
Jacinto 55C2
Jackson 47K3
Jackson 47I5
Jackson 47J4
Jacksonville 47K5
Jacksonville 47I4
Jacksonville 47L5
Jacmel 51J5
Jacobabad 26F4
Jacobina 53J6
Jacundá 53I4
Jaén 19E5
Jaffa, Cape 41H7
Jaffna 27G6
Jagdalpur 27H5
Jaguariaíva 55A4
Jaguaripe 55D1
Jahrom 26E4
Jaicós 53J5
Jaipur 27G4
Jaisalmer 27G4
Jakarta 29C8
Jakobstad 10M5
Jalālābād 27G3
Jalandhar 27G3
Jales 55A3
Jalgaon 27G4
Jalingo 32E4
Jalna 27G5
Jalpa 50D4
Jalpaiguri 27H4
Jamaica *country* 51I5
Jamaica Channel 51I5
Jambi 29C8
James *r.* 46H3
James Bay 45J4
Jamestown 41H6
Jamestown 46H2
Jamestown 48B1
Jammu 27G3
Jämsä 11N6
Jämsänkoski 10N6
Jamshedpur 27H4
Janaúba 55C1
Janesville 47J3
Jangheung 31B6
Janjanbureh 32B3
Januária 55B1
Japan *country* 31D5
Japan, Sea of 31D5
Japurá *r.* 52F4
Jaraguá 55A1
Jaraguá do Sul 55A4
Jardinópolis 55B3
Jarocin 13P5
Jarosław 23D6
Jarrettsville 48C3
Jaru 52F6
Järvenpää 11N6
Jasło 23D6
Jasper 44G4
Jasper 47I5
Jastrzębie-Zdrój 13Q6
Jászberény 21H1
Jataí 55A2
Jaú 55A3
Jauja 52C6
Java *i.* 29C8
Jawa, Laut *sea* 29D8
Jawhar 34E3
Jawor 13P5
Jaya, Puncak *mt.* 29F8
Jayapura 29G8
Jāzān 34E2
Jecheon 31C5
Jedburgh 16G5
Jeddah 34D1
Jefferson City 47I4
Jeffreys Bay 36G8
Jeju 31B6
Jēkabpils 11N8
Jelenia Góra 13O5
Jelgava 11M8
Jember 29D8
Jena 13M5
Jendouba 20C6
Jennings 47I5
Jeongeup 31B6
Jeonju 31B6
Jequié 55C1
Jequitinhonha 55C2
Jérémie 51J5
Jerez 50D4
Jerez de la Frontera 19C5
Jerome 46E3
Jersey *terr.* 15E9
Jersey City 48D2
Jerumenha 53J5
Jerusalem 33G1
Jervis Bay 42E5
Jervis Bay Territory *admin. div.* 42E5
Jesenice 20F1
Jesi 20E3
Jessheim 11G6
Jesup 47K5
Jhansi 27G4
Jiamusi 30C3
Ji'an 27K4
Ji'an 30B4
Jianyang 28D5
Jiaohe 30B4
Jiaxing 28E4
Jiayuguan 27I3
Jieznas 11N9
Jihlava 13O6
Jijel 32D1
Jijiga 34E3
Jilib 34E3
Jilin 30B4
Jilin *prov.* 30B4
Jīma 34D3
Jiménez 46G6
Jiménez 46H7
Jim Thorpe 48D2
Jinan 27K3
Jindo 31B6
Jindřichův Hradec 13O6
Jingdezhen 28D5
Jinggellic 42C5
Jingyu 30B4
Jingyuan 27J3
Jingzhou 27K3
Jinhae 31C6
Jinhua 28D5
Jining 27K3
Jinja 34D3
Jinotepe 51G6
Jinxi 28D5
Jinzhong 27K3
Jinzhou 27L2
Ji-Paraná 52F6
Jipijapa 52B4
Jiquiriçá 55D1
Jishou 27J4
Jiwen 30A2
Jixi 30C3
Jixian 30C3
Jizzax 26F2
Joaçaba 55A4
Joaíma 55C2
João Pessoa 53L5
João Pinheiro 55B2
Jodhpur 27G4
Joensuu 10P5
Jōetsu 31E5
Jõgeva 11O7
Johannesburg 37H4
John Day 46D3
John o' Groats 16F2
Johnsonburg 48B2
Johnstone 16E5
Johnstown 48D1
Johnstown 48B2
Johor Bahru 29C7
Jõhvi 11O7
Joinville 55A4
Joinville 18G2
Joliet 47J3
Jonava 11N9
Jonesboro 47I4
Jones Sound *sea chan.* 45J2
Jönköping 11I8
Jonquière 45K5
Joplin 47I4
Jordan *country* 33G1
Jorhat 27I4
Jos 32D4
José de San Martín 54B6
Joseph Bonaparte Gulf 40F2
Jos Plateau 32D4
Jouberton 37H4
Joutseno 11P6
Juan Aldama 46G7
Juan de Fuca Strait *strait* 44F5
Juazeiro 53J5
Juazeiro do Norte 53K5
Juba 33G4
Jubaland *area* 34E3
Jubba *r.* 34E4
Juchitán 50E5
Judenburg 13O7
Juigalpa 51G6
Juina 53G6
Juiz de Fora 55C3
Juliaca 52D7
Jumilla 19F4
Junagadh 27G4
Junction City 47H4
Jundiaí 55B3
Juneau 44E4
Junee 42C5
Jungfrau *mt.* 18H3
Junggar Pendi *basin* 27G2
Junín 54D4
Juodup 11N8
Jura *mts* 18G4
Jura *i.* 16D4
Jura, Sound of *sea chan.* 16D5
Jurbarkas 11M9
Jūrmala 11M8
Juruti 53G4
Jussara 55A1
Jutaí 52E5
Jutiapa 50G6
Juticalpa 51G6
Jutland *pen.* 11F8
Juventud, Isla de la *i.* 51H4
Jwaneng 36G3
Jyväskylä 10N5

K

K2 *mt.* 27G3
Kaarina 11M6
Kabale 34C4
Kabalo 35C4
Kabinda 35C4
Kabongo 35C4
Kābul 27F3
Kabwe 35C5
Kachchh, Rann of *marsh* 27G4
Kachia 32D4
K'ach'reti 23J8
Kachug 25L4
Kadapa 27G5
Kadıköy 21M4
Kadoma 35C5
Kadugli 33F3
Kaduna 32D3
Kaduy 22H4
Kadzherom 22L2
Kaédi 32B3
Kaélé 33E3
Kaesŏng 31B5
Kafanchan 32D4
Kafue 35C5
Kafue *r.* 35C5
Kaga 31E5
Kaga Bandoro 34B3
Kagoshima 31C7
Kagoshima *pref.* 31C7
Kaharlyk 23F6
Kahramanmaraş 26C3
Kaifeng 27K3
Kainan 31D6
Kairouan 20D7
Kaiyuan 30B4
Kaiyuan 27J4
Kajaani 10O4
Kakamega 34D3
Kakata 32B4
Kakhovka 21O1
Kakinada 27H5
Kakogawa 31D6
Kala 35D4
Kalaâ Kebira 20D7
Kalaburagi 27G5
Kalach 23I6
Kalach-na-Donu 23I6
Kalahari Desert 35C6
Kalajoki 10M4
Kalamaria 21J4
Kalamata 21J6
Kalamazoo 47J3
Kalanchak 21O1
Kalanshiyū ar Ramlī al Kabīr, Sarīr *des.* 33F1
Kale 21M6
Kalemie 35C4
Kalevala 10Q4
Kalgoorlie 40E6
Kali 20F2
Kaliningrad 11L9
Kalininsk 23J6
Kalininskaya 23H7
Kalinkavichy 23F5
Kalispell 46E2
Kalisz 13Q5
Kalix 10M4
Kalmar 11J8
Kaluga 23H5
Kalundborg 11G9
Kalush 23E6
Kalyazin 22H4
Kama 34C4
Kama *r.* 22L4
Kamaishi 31F5
Kambove 35C5
Kamchatka Peninsula 25Q4
Kamenka 23J5
Kamen'-na-Obi 24J4
Kamenolomni 23I7
Kamensk-Shakhtinskiy 23I6
Kamensk-Ural'skiy 24H4
Kamina 35C4
Kamloops 44F4
Kampala 34D3
Kampene 34C4
Kâmpóng Cham 29C6
Kâmpóng Spœ 29C6
Kâmpóng Thum 29C6
Kâmpôt 29C6
Kamuli 34D3
Kam"yanets'-Podil's'kyy 23E6
Kam"yanka-Buz'ka 23E6
Kamyanyets 11M10
Kāmyārān 33H1
Kamyshin 23J6
Kamyzyak 23K7
Kananga 35C4
Kanash 22J5
Kanazawa 31E5
Kanchipuram 27G5
Kandahār 26F3
Kandalaksha 10R3
Kandé 32D4
Kandi 32D3
Kandıra 21N4
Kandos 42D4
Kandy 27H6
Kandyagash 26E2
Kane'ohe 46□
Kanevskaya 23H7
Kangar 29C7
Kangaroo Island 41H7
Kangbyeon 30B4
Kangping 30A4
Kanin, Poluostrov *pen.* 22J2
Kankaanpää 11M6
Kankakee 47J3
Kankan 32C3
Kano 32D3
Kanoya 31C7
Kanpur 27H4
Kansas *state* 46H4
Kansas City 47I4
Kansk 25K4
Kantchari 32D3
Kantemirovka 23H6
Kanyamazane 37J3
Kanye 37G3
Kaolack 32B3
Kapiri Mposhi 35B5
Kapoeta 33G4
Kaposvár 20G1
Kapuskasing 45J5
Kapuvár 20G1
Kapyl' 11O10
Kara 32D4
Karabalyk 26F1
Karabük 23G8
Karacabey 21M4
Karacasu 21M6
Karachayevsk 23I8
Karachev 23G5
Karachi 26F4
Karaganda 27G2
Karagayly 27G2
Karahallı 21M5
Karaj 26E3
Kara-Köl 27G2
Karakol 27G2
Karakoram Range *mts* 27G3
Karakum Desert 26E2
Karakum Desert 26F3
Karaman 33G1
Karamanlı 21M6
Karamay 27H2
Karamea Bight *b.* 43C5
Karamürsel 21M4
!Karas *admin. reg.* 36C4
Karasburg 36D5
Kara Sea 24I2
Karasu 21N4
Karasuk 24I4
Karatau 27G2
Karatsu 31C6
Karbalā' 33H1
Karcag 21I1
Karditsa 21I5
Kärdla 11M7
Kardzhali 21K4
Kareima 33G3
Kareliya, Respublika *aut. rep.* 10R5
Karema 35D4
Kargalinskaya 23J8
Kargopol' 22H3
Kari 32E3
Kariba 35C5
Kariba, Lake *resr* 35C5
Kariba Dam 35C5
Karimata, Selat *strait* 29C8
Karlsi-Nuia 11N7
Karlivka 23G6
Karlovac 20F2
Karlovo 21K3
Karlovy Vary 13N5
Karlshamn 11I8
Karlskoga 11I7
Karlskrona 11I8
Karlsruhe 13L6
Karlstad 11H7
Karnal 27G4
Karnobat 21L3
Karoi 35C5
Karonga 35D4
Karpenisi 21I5
Karpogory 22J2
Karratha 40D4
Kars 23I8
Kärsava 11O8
Karsun 23J5
Kartal 21M4
Kartaly 24H4
Karvina 13Q6
Karwar 27G5
Karymskoye 25M4
Karystos 21K5
Kasaï, Plateau du 35C4
Kasama 35D5
Kasane 35C5
Kasese 34D3
Kāshān 26E3
Kashary 23I6
Kashi 27G3
Kashihara 31D6
Kashin 22H4
Kashira 23H5
Kashiwazaki 31E5
Kāshmar 26E3
Kashmir *reg.* 27G3
Kasimov 23I5
Kasongo 35C4
Kasongo-Lunda 35B4
Kaspiysk 23J8
Kassala 33G3
Kassel 13L5
Kasserine 20C7
Kastamonu 23G8
Kastoria 21I4
Kastsyukovichy 23G5
Kasulu 35D4
Kasungu 35D5
Katakwi 34D3
Katerini 21J4
Katete 35D5
Kathmandu 27H4
Kati 32C3
Katihar 27H4
Katima Mulilo 35C5
Katiola 32C4
Kati Thanda-Lake Eyre (North) 41H5
Kati Thanda-Lake Eyre (South) 41H5
Katlehong 37I4
Katni 27H4
Kato Achaïa 21I5
Katoomba 42E4
Katowice 13Q5
Kātrīnā, Jabal *mt.* 33G2
Katrine, Loch *l.* 16E4
Katrineholm 11J7
Katsina 32D3
Katsina-Ala 32D4
Katsuura 31F6
Kattegat *strait* 11G8
Kaua'i Channel 46□
Kauhajoki 10M5
Kauhava 10M5
Kaunas 11M9
Kaura-Namoda 32D3
Kavadarci 21J4
Kavak 23H8
Kavala 21K4
Kavalerovo 30D3
Kavarna 21M3
Kavīr, Dasht-e *des.* 26E3
Kawagoe 31E6
Kawaguchi 31E6
Kawasaki 31E6
Kawm Umbū 33G2
Kaya 32C3
Kayes 32B3
Kayseri 26C3
Kazakhstan *country* 26F2
Kazan' 22K5
Kazanlak 21K3
Kaz Dağı *mts* 21L5
Kazincbarcika 23D6
Keady 17F3
Kearney 46H3
Kebili 32D1
Keçiborlu 21N6
Kecskemét 21H1
Kdainiai 11M9
Kedong 30B3
Kędzierzyn-Koźle 13Q5
Keene 48E1
Keetmanshoop 36D4
Keffi 32D4
Keflavík 10□
Kegen 27G2
Kehra 11N7
Keighley 14F5
Keith 41I7
Keith 16G3
Kelibia 20D6
Kelkit *r.* 26C2
Kelm 11M9
Kélo 33E4
Kelowna 44G5
Keluang 29C7
Kem' 22G2
Kemalpaşa 21L5
Kemer 21N6
Kemerovo 24J4
Kemi 10N4
Kemijärvi 10O3
Kemnay 16G3
Kempele 10N4
Kempten (Allgäu) 13M7
Kempton Park 37I4
Kendal 14E4
Kendari 29E8
Kendawangan 29D8
Kenema 32B4
Kenge 35B4
Kenhardt 36E5
Kenitra 32C1
Kenmare 17C6
Kennewick 46D2
Kenosha 47J3
Kent 48A2
Kentucky *state* 47K4
Kenya *country* 34D3
Kenya, Mount 34D4
Keokuk 47I3
Keppel Bay 41K4
Kepsut 21M5
Kerang 42A5
Kerava 11N6
Kerba 19G5
Kerch 23H7
Kerema 38E2
Keren 33G3
Kerewan 32B3
Kericho 34D4
Kerkyra 21H5
Kermadec Islands 39I5
Kermān 26E3
Kermānshāh 33H1
Kermit 46G5
Kerrville 46H5
Kerteminde 11G9
Keşan 21L4
Keşap 23H8
Kesennuma 31F5
Keshan 30B2
Kesten'ga 10Q4
Keswick 14D4
Keszthely 20G1
Ketapang 29D8
Kettering 15G6
Keuruu 10N5
Keyihe 30A2
Key Largo 47K6
Keynsham 15E7
Keyser 48B3
Key West 47K7
Kgalagadi *admin. dist.* 36E3
Kgalagadi Transfrontier Park 35C6
Kgatleng *admin. dist.* 37H3
Kgotsong 37H4
Khabarovsk 30D2
Khabarovskiy Kray *admin. div.* 30D2
Khairpur 27F4
Khamar-Daban, Khrebet *mts* 25L4
Khamgaon 27G4
Khamis Mushayț 34E2
Khandyga 25O3
Khanka, Lake 30D3
Khanpur 27G4
Khanty-Mansiysk 24H3
Kharabali 23J7
Kharagpur 27H4
Khārijah, Wāḥāt al *oasis* 33G2
Kharkiv 23H6
Kharovsk 22I4
Khartoum 33G3
Khasavyurt 23J8
Khāsh 26F4
Khashuri 23I8
Khaybar 34D1
Khayelitsha 36D8
Khemis Miliana 19H5
Khenchela 20B7
Khenifra 32C1
Kherson 21O1
Khilok 25M4
Khlevnoye 23H5
Khmel'nyts'kyy 23E6
Khmil'nyk 23E6
Khomas *admin. reg.* 36C2
Khon Kaen 29C6
Khor 30D3
Khorol 30D3
Khorol 23G6
Khorramābād 33H1
Khorramshahr 33H1
Khorugh 27G3
Khouribga 32C1
Khromtau 26E1
Khrystynivka 23F6
Khŭjand 27F2
Khulays 34D1
Khulna 27H4
Khuma 37H4
Khuraysۜ 34E1
Klerksdorp 37H4
Khust 23D6
Khutsong 37H4
Khvalynsk 23K5
Khvoynaya 22G4
Khyber Pass 27G3
Kiama 42E5
Kibaha 35D4
Kiboga 34D3
Kibre Mengist 33G4
Kibungo 34D4
Kičevo 21I4
Kichmengskiy Gorodok 22J4
Kidal 32D3
Kidderminster 15E6
Kidsgrove 15E5
Kiel 13M3
Kielce 13R5
Kielder Water *resr* 14E3
Kieler Bucht *b.* 13M3
Kiev 23F6
Kiffa 32B3
Kifisia 21J5
Kigali 34D4
Kigoma 35C4
Kikinda 21I2
Kiknur 22J4
Kikwit 35B4
Kilchu 30C4
Kilcoole 17F4
Kilcoy 42F1
Kildare 17F4
Kil'dinstroy 10R2
Kilimanjaro *vol.* 34D4
Kilingi-Nõmme 11N7
Kiliya 21M2
Kilkeel 17G3
Kilkenny 17E5
Kilkis 21J4
Killarney 17C5
Killeen 46H5
Kilmarnock 16E5
Kil'mez' 22K4
Kilmore 42B6
Kilosa 35D4
Kilwinning 16E5
Kimbe 38F2
Kimberley 36G5
Kimberley Plateau 40F3
Kimch'aek 31C4
Kimjŏngsuk 30B4
Kimovsk 23H5
Kimpese 35B4
Kimry 22H4
Kinabalu, Gunung *mt.* 29D7
Kincardine 16F4
Kindersley 46F1
Kindia 32B3
Kindu 34C4
Kinel' 23K5
Kineshma 22I4
King Abdullah Economic City 26C4
Kingisepp 11P7
King Island 41I7
King Leopold Ranges *hills* 40E3
Kingman 49E3
King's Lynn 15H6
King Sound *b.* 40E3
Kingsport 47K4
Kingston 45K5
Kingston 51I5
Kingston 39G4
Kingston 48D2
Kingston 48D2
Kingston upon Hull 14G5
Kingstown 51L6
Kingsville 46H6
Kington 15D6
King William Island 45I3
King William's Town 37H7
Kinloss 16F3
Kinna 11H8
Kinross 16F4
Kinshasa 35B4
Kinston 47L4
Kintore 16G3
Kintyre *pen.* 16D5
Kipushi 35C5
Kirakira 39G3
Kirensk 25L4
Kireyevsk 23H5
Kiribati *country* 39J2
Kırıkkale 26C3
Kirillov 22H4
Kirishi 22G4
Kırkağaç 21L5
Kirkby 14E5
Kirkby in Ashfield 15F5
Kirkcaldy 16F4
Kirkcudbright 16E6
Kirkenes 10Q2
Kirkintilloch 16E5
Kirkkonummi 11N6
Kirkland Lake 45J5
Kırklareli 21L4
Kirksville 47I3
Kirkūk 33H1
Kirkwall 16G2
Kirov 23G5
Kirov 22K4
Kirovo-Chepetsk 22K4
Kirovohrad 23G6
Kirovsk 22F4
Kirovsk 10R3
Kirovs'ke 23G7
Kirovskiy 30D3
Kirs 22L4
Kirsanov 23I5
Kiruna 10L3
Kiryū 31E5
Kisangani 34C3
Kiselevsk 24J4
Kishi 32D4
Kishkenekol' 27G1
Kisii 34D4
Kiskunfélegyháza 21H1
Kiskunhalas 21H1
Kislovodsk 23I8
Kismaayo 34E4
Kisoro 33F5
Kissamos 21J7
Kissidougou 32B4
Kissimmee 47K6
Kisumu 34D4
Kita 32C3
Kitaibaraki 31F5
Kitakami 31F5
Kita-Kyūshū 31C6
Kitale 34D3
Kitami 30F4
Kitchener 48A1
Kitee 10Q5
Kitgum 34D3
Kittanning 48B2
Kitwe 35C5
Kiuruvesi 10O5
Kivu, Lac 34C4
Kizel 24G4
Kızılyurt 23J8
Kizlyar 23J8
Kizner 22K4
Kladno 13O5
Klagenfurt am Wörthersee 13O7
Klaipėda 11L9
Klaksvík 10□
Klamath *r.* 44F5
Klamath Falls 46C3
Klatovy 13N6
Klerksdorp 37H4
Kletnya 23G5
Kletskaya 23I6
Klimovo 23G5
Klin 22H4
Klintsy 23G5
Ključ 20G2
Kłodzko 13P5
Klosterneuburg 13P6
Kluczbork 13Q5
Klyetsk 11O10
Klyuchevskaya Sopka, Vulkan *vol.* 25R4
Knaresborough 14F4
Knighton 15D6
Knittelfeld 13O7
Knjaževac 21J3
Knockmealdown Mountains *hills* 17D5
Knoxville 47K4
Knysna 36F8
Kōbe 31D6
Koblenz 13K5
Kobryn 11N10
Kobuleti 23I8
Kočani 21J4
Kočevje 20F2
Kochi 27G6
Kōchi 31D6
Kochubeyevskoye 23I7
Kodiak 44C4
Kodiak Island 44C4
Kodino 22H3
Kodyma 23F6
Koforidua 32C4
Kōfu 31E6
Køge 11H9
Kohila 11N7
Kohima 27I4
Kohtla-Järve 11O7
Koidu-Sefadu 32B4
Kokkola 10M5
Kokomo 47J3
Kokosi 37H4
Kokpekty 27H2
Kokshetau 27F1
Kokstad 37I6
Kola 10R2
Kolaka 29E8
Kola Peninsula 22H2
Kol'chugino 22H4
Kolda 32B3
Kolding 11F9
Koléa 19H5
Kolhapur 27G5
Kolkata 27H4
Kollam 27G6
Kołobrzeg 13O3
Kolokani 32C3
Kolomna 23H5
Kolomyya 23E6
Kolondiéba 32C3
Kolonedale 38C2
Kolpashevo 24J4
Koluli 33H3
Kolwezi 35C5
Kolyma *r.* 25R3
Kolymskoye Nagor'ye *mts* 25R3
Kolyshley 23J5
Komaki 31E6
Komárno 13Q7
Komatsu 31E5
Kominternivs'ke 21N1
Komiža 20G3
Komló 20H1
Komotini 21K4
Komsomol's'k 23G6
Komsomol'sk-na-Amure 30E2
Kondoa 35D4
Kondopoga 22G3
Kondrovo 23G5
Kong Christian X Land *reg.* 45P2
Kong Frederik IX Land *reg.* 45M3
Kongolo 35C4
Kongoussi 32C3
Kongsberg 11F7
Kongsvinger 11H6
Konin 13Q4
Konosha 22I3
Konotop 23G6
Konstantinovka 30B2
Konstanz 13L7
Konya 33G1
Koper 20E2
Köping 11J7
Koprivnica 20G1
Korablino 23I5
Korçë 21I4
Korčula 20G3
Korčula *i.* 20G3
Korea Bay *g.* 31B5
Korea Strait *strait* 31C6
Korenovsk 23H7

Korets' 23E6
Körfez 21M4
Korhogo 32C4
Kōriyama 31F5
Korkuteli 21N6
Korla 27H2
Körmend 20G1
Koro i. 39H3
Koro 32C3
Köroğlu Dağları mts 21O4
Korogwe 35D4
Koror 29F7
Korosten' 23F6
Korostyshiv 23F6
Korsakov 30F3
Korsør 11G9
Korsun'-Shevchenkivs'kyy 23F6
Kortrijk 12I5
Koryazhma 22J3
Kos i. 21L6
Kosan 31B5
Kościan 13P4
Kosciuszko, Mount 42D6
Koshki 23K5
Košice 23D6
Kosŏng 31C5
Kosovo country 21I3
Kostanay 26F1
Kostenets 21J3
Kostinbrod 21J3
Kostomuksha 10Q4
Kostopil' 23E6
Kostroma 22I4
Kostyantynivka 23H6
Koszalin 13P3
Kszeg 20G1
Kota 27G4
Kotabaru 29D8
Kota Bharu 29C7
Kota Kinabalu 29D7
Kotel'nich 22K4
Kotel'nikovo 23I7
Kotel'nyy, Ostrov i. 25O2
Kotido 33G4
Kotka 11O6
Kotlas 22J3
Kotovo 23J6
Kotovsk 23I5
Koudougou 32C3
Koulikoro 32C3
Koumac 39G4
Koundâra 32B3
Koupéla 32C3
Kourou 53H2
Kouroussa 32C3
Kousséri 33E3
Koutiala 32C3
Kouvola 11O6
Kovdor 10Q3
Kovel' 23E6
Kovernino 22I4
Kovrov 22I4
Kovylkino 23I5
Kowanyama 41I3
Köyceğiz 21M6
Kozani 21I4
Kozara mts 20G2
Kozelets' 23F6
Kozel'sk 23G5
Kozhikode 27G5
Kozlu 21N4
Koz'modem'yansk 22J4
Kožuf mts 21J4
Kozyatyn 23F6
Krabi 29B7
Krâchéh 29C6
Kragerø 11F7
Kragujevac 21I2
Kraków 13Q5
Kramators'k 23H6
Kramfors 10J5
Kranidi 21J6
Kranj 20F1
Krāslava 11O9
Krasnaya Gorbatka 22I5
Krasnoarmeysk 23J6
Krasnoarmiys'k 23H6
Krasnoborsk 22J3
Krasnodar 23H7
Krasnodon 23H6
Krasnogorodsk 11P8
Krasnogvardeyskoye 23I7
Krasnohrad 23G6
Krasnohvardiys'ke 23G7
Krasnoperekops'k 23G7
Krasnoslobodsk 23I5
Krasnoyarsk 24K4
Krasnyy 23F5
Krasnyye Baki 22J4
Krasnyy Kholm 22H4
Krasnyy Kut 23J6
Krasnyy Luch 23H6
Krasnyy Lyman 23H6
Krasnyy Yar 23K7
Krasyliv 23E6
Krefeld 13K5
Kremenchuk 23G6
Krems an der Donau 13O6
Kresttsy 22G4
Kretinga 11L9
Kribi 32D4
Kristiansand 11E7
Kristianstad 11I8
Kristiansund 10E5
Kristinehamn 11I7
Kritiko Pelagos sea 21K6
Krk i. 20F2
Krolevets' 23G6
Kronshtadt 11P7
Kroonstad 37H4
Kropotkin 23I7
Krosno 23D6
Krotoszyn 13P5
Krui 29C8
Krumovgrad 21K4
Krupki 23F5
Kruševac 21I3
Krychaw 23F5
Krymsk 23H7
Kryvyy Rih 23G7
Ksar Chellala 19H6
Ksar el Boukhari 19H6
Ksar el Kebir 19D6
Ksour Essaf 20D7
Kstovo 22J4
Kuala Lipis 29C7
Kuala Lumpur 29C7
Kuala Terengganu 29C7
Kuandian 30B4
Kuantan 29C7
Kubrat 21L3
Kuching 29D7
Kuçovë 21H4
Kudat 29D7
Kufstein 13N7
Kugesi 22J4
Kuhmo 10P4
Kuito 35B5
Kujang 31B5
Kuji 31F4
Kukës 21I3
Kukmor 22K4
Kula 21M5
Kular 25O2
Kuldīga 11L8
Kulebaki 23I5
Kulmbach 13M5
Kŭlob 27F3
Kul'sary 26E2
Kulunda 24I4
Kumagaya 31E5
Kumamoto 31C6
Kumano 31E6
Kumanovo 21I3
Kumasi 32C4
Kumba 32D4
Kumdah 34E1
Kumeny 22K4
Kumertau 24G4
Kumi 33G4
Kumla 11I7
Kumo 32E3
Kumylzhenskiy 23I6
Kungälv 11G8
Kungsbacka 11H8
Kunlun Shan mts 27G3
Kunming 27J4
Kuopio 10O5
Kupang 40E2
Kupiškis 11N9
Kup''yans'k 23H6
Kuqa 27H2
Kurashiki 31D6
Kurayoshi 31D6
Kurchatov 23G6
Kurdistan reg. 26D3
Kure 31D6
Kuressaare 11M7
Kurgan 24H4
Kurganinsk 23I7
Kurikka 10M5
Kuril Islands 30H3
Kurkino 23H5
Kurmuk 33G3
Kurnool 27G5
Kurri Kurri 42E4
Kursavka 23I7
Kursk 23H6
Kurskaya 23J7
Kurşunlu 23G8
Kuruman 36F4
Kurume 31C6
Kurunegala 27H6
Kuşadası 21L6
Kushchëvskaya 23H7
Kushiro 30G4
Kusmuryn 26F1
Kusŏng 31B5
Kütahya 21M5
Kutaisi 23I8
Kutjevo 20G2
Kutno 13Q4
Kutu 34B4
Kutztown 48D2
Kuusamo 10P4
Kuusankoski 11O6
Kuvshinovo 22G4
Kuwait country 26D4
Kuwait 26D4
Kuybyshev 24I4
Kuybysheve 23H7
Kuybyshevskoye Vodokhranilishche resr 23K5
Kuytun 27H2
Kuyucak 21M6
Kuznetsk 23J5
Kuznetsovs'k 23E6
Kuzovatovo 23J5
Kvarnerić sea chan. 20F2
KwaDukuza 37J5
Kwale 32D4
KwaMashu 37J5
Kwa Mtoro 35D4
KwaNobuhle 37G7
Kwatarkwashi 32D3
Kwatinidubu 37H7
KwaZulu-Natal prov. 37J5
Kwekwe 35C5
Kweneng admin. dist. 36G2
Kwidzyn 13Q4
Kyakhta 25L4
Kyaukpyu 27I5
Kymi 21K5
Kyneton 42B6
Kyogle 42F2
Kyōto 31D6
Kyparissia 21I6
Kyrgyzstan country 27G2
Kythira i. 21J6
Kyūshū i. 31C7
Kyustendil 21J3
Kyzyl 24K4
Kyzylkum Desert 26F2
Kyzyl-Mazhalyk 24K4
Kyzylorda 26F2

L

Laagri 11N7
Laâyoune 32B2
La Banda 54D3
Labasa 39H3
Labé 32B3
Labinsk 23I7
Labouheyre 18D4
Laboulaye 54D4
Labrador reg. 45L4
Labrador City 45L4
Labrador Sea 45M3
Lábrea 52F5
Labuhanbilik 27J6
Labuna 29E8
Labytnangi 24H3
La Carlota 54D4
Lac du Bonnet 47H1
La Ceiba 51G5
Lachlan r. 42A5
La Chorrera 51I7
Lachute 45K5
La Ciotat 18G5
Lac La Biche 44G4
Laconia 48F1
La Crosse 47I3
Ladainha 55C2
Ladik 23G8
Ladoga, Lake 11Q6
Ladysmith 37I5
Lae 38E2
Lafayette 47J3
Lafayette 47I5
Lafia 32D4
Lafiagi 32D4
La Flèche 18D3
Lagan' 23J7
Lagan r. 17G3
Lagarto 53K6
Lågen r. 11G7
Lages 55A4
Laghouat 32D1
Lagoa Santa 55C2
Lagoa Vermelha 55A5
Lagos 32D4
Lagos 19B5
Lagosa 35C4
La Grande 46D2
La Grande 4, Réservoir resr 45K4
La Grange 47J5
La Gran Sabana plat. 52F2
Laguna 55A5
Laha 30B2
La Habra 49D4
Lahad Datu 29D7
Lahat 29C8
Laḥij 34E2
Laholm 11H8
Lahore 27G3
Lahti 11N6
Laï 33E4
Laidley 42F1
Laihia 10M5
Laishevo 22K5
Laitila 11L6
Laiyang 28E4
Laizhou Wan b. 27K3
Lajeado 55A5
Lajes 53K5
La Junta 46G4
Lake Cargelligo 42C4
Lake Charles 47I5
Lake City 47K5
Lake Elsinore 49D4
Lake Havasu City 49E3
Lakehurst 48D2
Lakeland 47K6
Lake Providence 47I5
Lakes Entrance 42D6
Lakeside 48C4
Lakewood 48D2
Lakewood 48B1
Lakhdenpokh'ya 10Q6
Lakota 32C4
Laksefjorden sea chan. 10O1
Lakshadweep 27G5
La Ligua 54B4
Lalín 19B2
La Línea de la Concepción 19D5
Lalitpur 27G4
La Louvière 12J5
Lamar 46G4
La Martre, Lac l. 44G3
Lambaréné 34B4
Lambayeque 52C5
Lambert's Bay 36D7
Lambeth 48A1
Lamego 19C3
La Merced 54C3
La Merced 52C6
La Mesa 49D4
Lamesa 46G5
Lamia 21J5
Lammermuir Hills 16G5
Lamont 49C3
Lampang 29B6
Lampazos 46G6
Lamu 34E4
Lancaster 14E4
Lancaster 49C3
Lancaster 48C2
Lancaster 47K5
Lancaster Sound strait 45J2
Landeck 13M7
Lander 46F3
Landsberg am Lech 13M6
Land's End pt 15B8
Landshut 13N6
Landskrona 11H9
Langenthal 18H3
Langjökull ice cap 10□2
Langport 15E7
Langres 18G3
Langsa 27I6
Långsele 10J5
Lannion 18C2
Lansing 47K3
Lanxi 30B3
Lanzarote i. 32B2
Lanzhou 27J3
Laoag City 29E6
Lao Cai 27J4
Laojunmiao 27I3
Laon 18F2
La Oroya 52C6
Laos country 29C6
Laotougou 30C4
Lapa 55A4
La Palma i. 32B2
La Palma 51I7
La Paz 54E4
La Paz 52E7
La Paz 50G6
La Paz 46E7
La Pérouse Strait strait 30F3
La Plata 54E4
La Plata, Río de sea chan. 54E4
La Pola Siero 19D2
Lappeenranta 11P6
Lappland reg. 10K3
Lâpseki 21L4
Laptev Sea 25N2
Lapua 10M5
La Quiaca 54C2
L'Aquila 20E3
La Quinta 49D4
Larache 19C6
Laramie 46F3
Laranjal Paulista 55B3
Laranjeiras do Sul 54F3
Larba 19H5
Laredo 46H6
Largs 16E5
L'Ariana 20D6
La Rioja 54C3
La Rioja aut. comm. 19E2
Larisa 21J5
Larne 17G3
La Rochelle 18D3
La Roche-sur-Yon 18D3
La Romana 51K5
La Ronge 44H4
La Ronge, Lac l. 44H4
Larvik 11G7
Las Cruces 46F5
La Serena 54B3
Las Flores 54E5
Las Heras 54C4
Las Palmas de Gran Canaria 32B2
La Spezia 20C2
Las Tablas 51H7
Las Tunas 51I4
Las Varas 50C4
Las Varillas 54D4
Las Vegas 46F4
Las Vegas 49E2
Latacunga 52C4
Latakia 33G1
La Teste-de-Buch 18D4
Latina 20E4
La Tuque 47M2
Latvia country 11N8
Lauchhammer 13N5
Laudio 19E2
Launceston 41J8
Launceston 15C8
Laurel 47J5
Laureldale 48D2
Laurel Hill hills 48B3
Laurinburg 47L5
Lausanne 18H3
Lautoka 39H3
Laval 18D2
La Vall d'Uixó 19F4
La Vega 51J5
Lavras 55B3
Lawra 32C3
Lawrence 47H4
Lawrence 48F1
Lawrenceburg 47J4
Lawton 46H5
Lazarev 30F1
Lazarevac 21I2
Lázaro Cárdenas 50D5
Lazdijai 11M9
Leatherhead 15G7
Lebanon country 33G1
Lebanon 47I4
Lebanon 48E1
Lebanon 48C2
Lebedyan' 23H5
Lębork 13P3
Lebowakgomo 37I3
Lebrija 19C5
Lebu 54B5
Lecce 20H4
Lecco 20C2
Lechaina 21I6
Le Creusot 18G3
Le Crotoy 15I8
Ledesma 19D3
Ledmozero 10R4
Leeds 14F5
Leek 15E5
Leesburg 48C3
Leesville 47I5
Leeuwarden 13J4
Leeuwin, Cape 40D6
Leeward Islands 51L5
Lefkada 21I5
Lefkada i. 21I5
Lefkimmi 21I5
Legnago 20D2
Legnica 13P5
Le Havre 15H9
Lehmo 10P5
Leibnitz 13O7
Leicester 15F6
Leiden 12J4
Leigh 14E5
Leigh Creek 41H6
Leighton Buzzard 15G7
Leipzig 13N5
Leiria 19B4
Leirvik 11D7
Leizhou Bandao pen. 27J4
Le Kef 20C6
Leksand 11I6
Lelystad 12J4
Le Mans 18E2
Le Mars 47H3
Leme 55B3
Lemmon 46G2
Lemoore 49C2
Le Murge hills 20G4
Lemvig 11F8
Lena r. 25N2
Lenham 15H7
Lenine 23G7
Leningradskaya 23H7
Leningradskaya Oblast' admin. div. 11R7
Leningradskiy 25S3
Leninsk 23J6
Leninskiy 23H5
Leninsk-Kuznetskiy 24J4
Leninskoye 22J4
Leninskoye 30D3
Lens 18F1
Lensk 25M3
Lenti 20G1
Lentini 20F6
Léo 32C3
Leoben 13O7
Leominster 15E6
Leominster 48F1
León 50D4
León 51G6
León 19D2
Leongatha 42B7
Leonidio 21J6
Leonidovo 30F2
Leonora 40E5
Leopoldina 55C3
Lepontine Alps mts 18I3
Le Puy-en-Velay 18F4
Lerala 37H2
Léré 32C3
Lerma 19E2
Le Roy 48C1
Lerum 11H8
Lerwick 16□
Lesbos i. 21K5
Les Cayes 51J5
Leshan 27J4
Leshukonskoye 22J2
Leskovac 21I3
Lesosibirsk 24K4
Lesotho country 37I5
Lesozavodsk 30D3
L'Espérance Rock i. 39I5
Les Sables-d'Olonne 18D3
Lesser Antilles is 51K6
Lesser Caucasus mts 23I8
Lesser Slave Lake 44G4
Leszno 13P5
Letchworth Garden City 15G7
Lethbridge 44G5
Leticia 52E4
Letnerechenskiy 22G2
Le Touquet-Paris-Plage 15I8
Letterkenny 17E3
Leuchars 16G4
Leuven 12J5
Levanger 10G5
Levashi 23J8
Levelland 46G5
Leven 14G5
Leven, Loch l. 16F4
Lévêque, Cape 40E3
Leverkusen 13K5
Levice 13Q6
Levittown 48E2
Levittown 48D2
Lev Tolstoy 23H5
Lewes 15H8
Lewis, Isle of i. 16C2
Lewisburg 48C2
Lewis Range mts 46E2
Lewiston 46D2
Lewistown 46F2
Lewistown 48C2
Lexington 47K4
Lexington 46H3
Lexington 48B4
Lezhë 21H4
L'gov 23G6
Lhasa 27I4
Lianyungang 28D4
Liaodong Wan b. 27L2
Liaoning prov. 30A4
Liaoyang 30A4
Liaoyuan 30B4
Liaozhong 30A4
Libenge 34B3
Liberal 46G4
Liberec 13O5
Liberia country 32C4
Liberia 51G6
Libourne 18D4
Libreville 34A3
Libya country 33E2
Libyan Desert 33F2
Libyan Plateau 33F1
Licata 20E6
Lichfield 15F6
Lichinga 35D5
Lichtenburg 37H4
Lida 11N10
Lidköping 11H7
Liebig, Mount 40G4
Liechtenstein country 18I3
Liège 13J5
Lieksa 10Q5
Lienz 13N7
Liepāja 11L8
Liezen 13O7
Liffey r. 17F4
Lifford 17E3
Lightning Ridge 42C2
Ligurian Sea 18I5
Lika reg. 20F2
Likasi 35C5
Likhoslavl' 22G4
Lilla Edet 11H7
Lille 18F1
Lillehammer 11G6
Lillestrøm 11G7
Lilongwe 35D5
Lima 52C6
Lima 48C1
Lima 47K3
Lima Duarte 55C3
Liman 23J7
Limassol 33G1
Limavady 17F2
Limbaži 11N8
Limeira 55B3
Limerick 17D5
Limmen Bight b. 41H2
Limnos i. 21K5
Limoeiro 53K5
Limoges 18E4
Limon 46G4
Limoux 18F5
Limpopo prov. 37I2
Limpopo r. 37K3
Limpopo, Parque Nacional do 37J2
Linares 54B5
Linares 46H7
Linares 19E4
Lincoln 54D4
Lincoln 14G5
Lincoln 47J3
Lincoln 47H3
Lindau (Bodensee) 13L7
Linden 53G2
Lindi 35D4
Lindian 30B3
Line Islands 6
Linfen 27K3
Lingen (Ems) 13K4
Lingga, Kepulauan is 29C8
Linhares 55C2
Linjiang 30B4
Linköping 11I7
Linkou 30C3
Linlithgow 16F5
Linnhe, Loch inlet 16D4
Lins 55A3
Linxi 27K2
Linxia 27J3
Linyi 27K3
Linz 13O6
Lion, Golfe du g. 18F5
Lipetsk 23H5
Lipova 21I1
Lira 34D3
Lisala 34C3
Lisbon 19B4
Lisburn 17F3
Lishu 30B4
Lisieux 18E2
Liski 23H6
Lismore 42F2
Lithgow 42E4
Lithuania country 11M9
Litoměřice 13O5
Little Andaman i. 27I5
Little Belt sea chan. 11F9
Little Cayman i. 51H5
Little Falls 47I2
Littlefield 46G5
Littlehampton 15G8
Little Minch sea chan. 16B3
Little Rock 47I5
Liuhe 30B4
Liupanshui 27J4
Liuzhou 27J4
Livadeia 21J5
Livermore 49B2
Liverpool 42E4
Liverpool 14E5
Liverpool Plains 42E3
Liverpool Range mts 42D3
Livingston 16F5
Livingston 46E2
Livingston 47I5
Livingstone 35C5
Livno 20G3
Livny 23H5
Livorno 20D3
Livramento de Nossa Senhora 55C1
Lizard Point 15B9
Ljubljana 20F1
Ljungby 11H8
Ljusdal 11J6
Llandovery 15D7
Llandudno 14D5
Llanelli 15C7
Llangollen 15D6
Llano Estacado plain 46G5
Llanos plain 52E2
Llantrisant 15D7
Llay 15D5
Lleida 19G3
Lobamba 37J4
Lobatse 37G3
Loberia 54E5
Lobito 35B5
Lobos 54E5
Lochy, Loch l. 16E4
Lockerbie 16F5
Lock Haven 48C2
Lockport 48B1
Lodeynoye Pole 22G3
Lodi 20C2
Lodi 49B1
Lodja 34C4
Łódź 13Q5
Lofoten is 10H2
Log 23I6
Logan 46E3
Logan, Mount 44D3
Logatec 20F2
Logroño 19E2
Loimaa 11M6
Loire r. 18C3
Loja 52C4
Loja 19D5
Løkken 10F5
Loknya 22F4
Lokoja 32D4
Lokossa 32D4
Lokot' 23G5
Lolland i. 11G9
Lom 21J3
Lomas de Zamora 54E4
Lombok i. 40D1
Lombok, Selat sea chan. 29D8
Lomé 32D4
Lomond, Loch l. 16E4
Lomonosov 11P7
Lompoc 49B3
Łomża 13S4
London 48A1
London 15G7
London 47K4
Londonderry 17E3
Londonderry, Cape 40F2
Londrina 55A3
Longa, Proliv sea chan. 25S2
Long Ashton 15E7
Long Beach 49C4
Long Branch 48E2
Long Eaton 15F6
Longford 17E4
Long Island 48E2
Long Island Sound sea chan. 48E2
Longjiang 30A3
Longmeadow 48E1
Long Melford 15H6
Longmont 46F3
Long Point Bay 48A1
Longreach 41I4
Longtown 14E3
Longview 47I5
Long Xuyên 29C6
Longyan 28D5
Longyearbyen 24C2
Lönsboda 11I8
Lons-le-Saunier 18G3
Lop Buri 29C6
Lop Nur salt flat 27I2
Lorain 45J5
Lorca 19F5
Lord Howe Island 41L6
Lorena 55B3
Loreto 53I5
Lorient 18C3
Lorn, Firth of est. 16D4
Lorrain, Plateau 18H2
Los Alamos 46F4
Los Ángeles 54B5
Los Angeles 49C3
Los Banos 49B2
Los Juríes 54D3
Los Mochis 46F6
Lossiemouth 16F3
Los Teques 52E1
Lostwithiel 15C8
Los Vilos 54B4
Lot r. 18E4
Lota 54B5
Louangnamtha 28C5
Louangphabang 29C6
Louga 32B3
Loughborough 15F6
Loughrea 17D4
Loughton 15H7
Louisiade Archipelago is 41K2
Louisiana state 47I5
Louisville 47J4
Loukhi 10R3
Loulé 19B5
Loum 32D4
Louny 13N5
Lourdes 18D5
Louth 14G5
Loutra Aidipsou 21J5
Lovech 21K3
Loviisa 11O6
Lovington 46G5
Lowell 48F1
Lower Hutt 43E5
Lower Lough Erne l. 17E3
Lowestoft 15I6
Łowicz 13Q4
Loxton 41I6
Loyauté, Îles is 39G4
Loyew 23F6
Loznica 21H2
Lozova 23H6
Luanda 35B4
Luanshya 35C5
Luau 35C5
Luba 32D4
Lubaczów 23D6
Lubango 35B5
Lubao 35C4
Lubartów 23D6
Lubbock 46G5
Lübeck 13M4
Lubin 13P5
Lublin 23D6
Lubny 23G6
Lubumbashi 35C5
Lucala 35B4
Lucapa 35C4
Lucca 20D3
Luce Bay 16E6
Lucélia 55A3
Lucena 29E6
Lucena 19D5
Lučenec 13Q6
Lucera 20F4
Lucerne 18I3
Luchegorsk 30D3
Luckenwalde 13N4
Lucknow 27H4
Lüdenscheid 13K5
Lüderitz 36B4
Ludhiana 27G3
Ludvika 11I6
Ludwigsburg 13L6
Ludwigshafen am Rhein 13L6
Ludza 11O8
Luebo 35C4
Luena 35B5
Lufkin 47I5
Luga 11P7
Lugano 18I3
Lugo 20D2
Lugo 19C2
Lugoj 21I2
Luhans'k 23H6
Lukhovitsy 23H5
Lukovit 21K3
Łuków 23D6
Lukoyanov 23J5
Luleå 10M4
Luleälven r. 10M4
Lüleburgaz 21L4
Lumberton 47L5
Lumbrales 19C3
Lumezzane 20D2
Lund 11H9
Lundy i. 15C7
Lune r. 14E4
Lüneburg 13M4
Lunéville 18H2
Luninyets 11O10
Lunsar 32B4
Luobei 30C3
Luohe 27K3
Luoyang 27K3
Lupane 35C5
Lupeni 21J2
Lurgan 17F3
Lusaka 35C5
Lusambo 35C4
Lushnjë 21H4
Lūt, Kavīr-e des. 26E3
Luton 15G7
Luts'k 23E6
Lutzville 36D6
Luwero 34D3
Luwuk 29E8
Luxembourg country 13K6
Luxembourg 13K6
Luxor 33G2
Luza 22J3
Luzhou 27J4
Luziânia 55B2
Luzon i. 29E6
Luzon Strait strait 29E5
L'viv 23E6
Lyakhavichy 11O10
Lycksele 10K4
Lydd 15H8
Lyel'chytsy 23F6
Lyepyel' 11P9
Lyme Bay 15E8
Lyme Regis 15E8
Lymington 15F8
Lynchburg 48B4
Lynn 48F1
Lyon 18G4
Lyozna 23F5
Lysekil 11G7
Lyskovo 22J4
Lys'va 24G4
Lysychans'k 23H6
Lysyye Gory 23J6
Lytham St Anne's 14D5
Lyuban' 11P10
Lyubim 22I4
Lyudinovo 23G5

M

Ma'ān 33G1
Maastricht 13J5
Mabaruma 52G2
Mablethorpe 14H5
Mabopane 37I3
Macaé 55C3
Macajuba 55C1
Macao 27K4
Macapá 53H3
Macará 52C4
Macarani 55C1
Macas 52C4
Macau 53K5
Macclesfield 14E5
MacDonnell Ranges mts 40G4
Macduff 16G3
Macedonia country see North Macedonia
Maceió 53K5
Macenta 32C4
Macerata 20E3
Machachi 52C4
Machakos 34D4
Machala 52C4
Machilipatnam 27H5
Machiques 52D1
Machu Picchu tourist site 52D6
Machynlleth 15D6
Măcin 21M2
Macina 32C3
Macintyre Brook r. 42E2
Maçka 23H8
Mackay 41J4
Mackenzie r. 44E3
Mackenzie Bay 44E3
Mackenzie Mountains 44E3
Macksville 42F3
Maclean 42F2
Macomb 47I3
Mâcon 18G3
Macon 47K5
Macon 47I4
Madadeni 37J4
Madagascar country 35E6
Madan 21K4
Madang 38E2
Madaoua 32D3
Madeira r. 52G4
Madeira terr. 32B1
Madera 46F6
Madingou 35B4
Madison 47J4
Madison 47H3
Madison 47J3
Madison Heights 48B4
Madisonville 47J4
Madona 11O8
Madra Dağı mts 21L5
Madrakah 33G2
Madre, Laguna lag. 47H6
Madre del Sur, Sierra mts 50D5
Madre Occidental, Sierra mts 46F6
Madre Oriental, Sierra mts 46G6
Madrid 19E3
Madurai 27G6
Maebashi 31E5
Maevatanana 35E5
Mafadi mt. 37I5
Mafeteng 37H5
Maffra 42C6
Mafinga 35D4
Mafra 55A4
Magabeni 37J6
Magadan 25Q4
Magangué 52D2
Magaria 32D3
Magas 23J8
Magdagachi 30B1
Magdalena 46E5
Magdeburg 13M4
Magellan, Strait of 54B8
Maggiore, Lake 20C2
Magherafelt 17F3
Maghnia 19F6
Maghull 14E5
Magnitogorsk 24G4
Magnolia 47I5
Mago 30F1
Magṭa' Laḥjar 32B3
Magwe 27I4
Mahābād 33H1
Mahajanga 35E5
Mahalapye 37H2
Mahalevona 35E5
Mahanoro 35E5
Maha Sarakham 27J5
Mahd adh Dhahab 34E1
Mahdia 19G6
Mahdia 53G2
Mahdia 20D7
Mahenge 35D4
Mahilyow 23F5
Maicao 52D1
Maidenhead 15G7
Maidstone 15H7
Maiduguri 32E3
Maīmanah 26F3
Maine state 47N2
Maine, Gulf of 45L5
Mainland i. 16F1
Mainland i. 16□
Maintirano 35E5
Mainz 13L5
Maitland 42E4
Maizuru 31D6
Majene 29D8
Majorca i. 19H4
Makabana 34B4
Makale 29D8
Makanshy 27H2
Makarov 30F2
Makar'yev 22I4
Makassar 29D8
Makassar, Selat strait 29D8
Makat 26E2
Makeni 32B4
Makgadikgadi depr. 35C6
Makhachkala 23J8
Makinsk 27G1
Makiyivka 23H6
Makó 21I1
Makokou 34B3
Maksatikha 22G4
Makunguwiro 35D5
Makurdi 32D4
Malabar Coast 27G5
Malabo 32D4
Malacca, Strait of strait 29B7
Maladzyechna 11O9
Málaga 19D5
Malaita i. 39G2
Malakal 33G4
Malang 29D8
Malanje 35B4
Mälaren l. 11J7
Malargüe 54C5
Malatya 26C3
Malawi country 35D5
Malaya Vishera 22G4
Malāyer 33H1
Malaysia country 29C7
Malbork 13Q3
Maldives country 27G6
Maldon 42B6
Maldon 15H7
Maldonado 54F4
Male 7
Malgobek 23J8
Malhada 55C1
Mali country 32C3
Malili 38C2
Malindi 34E4
Malin Head 17E2
Malkara 21L4
Mallawī 33G2
Mallet 55A4
Mallow 17D5
Malmedy 13K5
Malmesbury 36D7
Malmesbury 15E7
Malmö 11H9
Malmyzh 22K4
Maloshuyka 22H3
Maloyaroslavets 23H5
Malta country 20F7
Malta 11O8
Malta Channel 20F6
Maltby 14F5
Maluku, Laut sea 29E8
Malung 11H6
Maluti Mountains 37I5
Malvern 47I5
Malyn 23F6
Malyye Derbety 23J7
Mamadysh 22K5
Mamelodi 37I3
Mamfe 32D4
Mamoré r. 52E6
Mamou 32B3
Mamuju 38B2
Man 32C4
Man, Isle of terr. 14C4
Manacapuru 52F4
Manacor 19H4
Manado 29E7
Managua 51G6
Manakara 35E6
Manama 34F1
Mananara Avaratra 35E5
Mananjary 35E6
Manassas 48C3
Manaus 52F4
Manavgat 33G1
Manchester 14E5
Manchester 48E2
Manchester 48C3
Manchester 48F1
Mandal 11E7
Mandalay 27I4
Mandalgovĭ 27J2
Mandan 46G2
Mandeville 51I5
Mandritsara 35E5
Mandsaur 27G4
Mandurah 40D6
Manduria 20G4
Mandya 27G5
Manevychi 23E6
Manfredonia 20F4
Manga 55C1
Manga 32C3
Mangai 34B4
Mangalia 21M3
Mangaluru 27G5
Mangaung 37H5
Mangotsfield 15E7
Mangualde 19C3
Manhattan 47H4
Manhica 37K3
Manhuaçu 55C3
Manicouagan, Réservoir resr 45L4
Maniitsoq 45M3
Manila 29E6
Manilla 42E3
Manisa 21L5
Manistee 47J3
Manitoba prov. 45I4
Manitoba, Lake 45I4
Manitowoc 47J3
Manizales 52C2
Manja 35E6
Mankato 47I3
Mankono 32C4
Mannar, Gulf of 27G6
Mannheim 13L6
Mannington 48A3
Manoel Ribas 55A4
Manoel Vitorino 55C1
Manp'o 30B4
Manresa 19G3
Mansa 35C5
Mansa Konko 32B3
Mansfield 42C6
Mansfield 15F5
Mansfield 47I5
Mansfield 47K3
Manta 52B4
Manteca 49B2
Mantena 55C2
Mantes-la-Jolie 18E2
Mantiqueira, Serra da mts 55B3
Mantoudi 21J5
Mäntsälä 11N6
Mänttä 10N5
Mantua 20D2
Manturovo 22J4
Manukau 43E3
Manyas 21L4
Manyoni 35D4
Manzanillo 51I4
Manzanillo 50D5
Manzhouli 27K2
Manzini 37J4
Mao 33E3
Maó 19I4
Maokeng 37H4
Maple Creek 46F2
Maputo 37K3
Maputo prov. 37K3
Maputsoe 37H5
Maqteïr reg. 32B2
Maquela do Zombo 35B4
Maraã 52E4
Marabá 53I5
Maracaibo 52D1
Maracaibo, Lake 52D2
Maracaju 54E2
Maracás 55C1
Maracay 52E1
Maradi 32D3
Marajó, Ilha de i. 53H4
Marand 26D3
Marañón r. 52D4
Maraú 55D1
Marbella 19D5
Marble Hall 37I3
Marburg 13L5
Marburg 37J6
Marcali 20G1
March 15H6
Marcona 52C7
Mardan 27G3
Mar del Plata 54E5
Mardin 33H1
Maree, Loch l. 16D3
Margao 27G5
Margate 15I7
Margherita Peak 34C3
Marg'ilon 27G2
Marhanets' 23G7
Mariana 55C3
Marianna 47I5
Marianna 47J5
Ma'rib 34E2
Maribor 20F1
Mariehamn 11K6
Mariental 36C3
Mariestad 11H7
Marietta 47K5
Marietta 48A3
Marignane 18G5
Mariinsk 24J4
Mariinskiy Posad 22J4
Marijampol 11M9
Marília 55A3
Marín 19B2
Marina 49B2
Mar''ina Horka 11P10
Marinette 47J2
Maringá 55A3
Marinha Grande 19B4
Marion 47J3
Marion 47K3
Marion 47L5
Marion 47K4
Maritime Alps mts 18H4
Mariupol' 23H7
Marīvān 33H1
Marka 34E3
Market Deeping 15G6
Market Harborough 15G6
Market Weighton 14G5
Markovo 25S3
Marks 23J6
Marmande 18E4
Marmara, Sea of g. 21M4
Marmaris 21M6
Marne r. 18F2
Marne-la-Vallée 18F2
Maroantsetra 35E5
Marondera 35D5
Maroochydore 42F1
Maroua 33E3
Marovoay 35E5
Marquesas Islands 6
Marquette 47J2
Marra, Jebel mt. 33F3
Marra, Jebel plat. 33F3
Marrakech 32C1
Marsá al 'Alam 33G2
Marsabit 34D3
Marsala 20E6
Marsá Maṭrūḥ 33F1
Marseille 18G5
Marshall 47H3
Marshall 47I4
Marshall 47I5
Marshall Islands country 6
Marshalltown 47I3
Märsta 11J7
Martapura 29D8
Martigny 18H3
Martin 13Q6
Martinho Campos 55B2
Martinique terr. 51L6
Martinsburg 48C3
Martinsville 47L4
Martok 26E1
Martos 19E5
Maruim 53K6
Mary 26F3
Maryborough 41K5
Maryland state 48C3
Marysville 49B1
Maryville 47I3
Maryville 47K4
Masai Steppe plain 35D4
Masaka 34D4
Masan 31C6
Masasi 35D5
Mascara 19G6
Mascote 55D1
Maseru 37H5
Mashhad 26E3
Masilo 37H5
Masindi 34D3
Masjed Soleymān 33H1
Mask, Lough l. 17C4
Mason City 47I3
Massa 20D2
Massachusetts state 48E1
Massachusetts Bay 48F1
Massafra 20G4
Massango 35B4
Massawa 33G3
Massenya 33E3
Massif Central mts 18F4
Massillon 48A2
Masterton 43E5
Mastung 26F4
Mastūrah 34D1
Masty 11N10
Masuda 31C6
Masvingo 35D6
Masvingo prov. 37J1
Matabeleland South prov. 37I1
Matadi 35B4
Matagalpa 51G6
Matam 32B3
Matamey 32D3
Matamoros 46G6
Matamoros 47H6
Matane 47N2
Matanzas 47K7
Mataram 29D8
Mataró 19H3
Matā'utu 39I3
Matehuala 50D4
Matemanga 35D5
Matera 20G4
Mateur 20C6
Mathura 27G4
Mati 29E7
Matias Cardoso 55C1
Matías Romero 50E5
Matlock 15F5
Mato Grosso state 55A1
Mato Grosso, Planalto do plat. 53H7
Matola 37K3
Matosinhos 19B3
Mato Verde 55C1
Matsue 31D6
Matsumoto 31E5
Matsuyama 31D6
Matterhorn mt. 18H4
Matterhorn mt. 46D3
Maturín 52F2
Matwabeng 37H5
Mauchline 16E5
Maui i. 46□
Maun 35C5
Maunatlala 37H2
Mauritania country 32B3
Mauritius country 7
Mavinga 35C5
Mawlamyaing 27I5
Mawqaq 33H2
Mayagüez 51K5
Maya Mountains 50G5
Maybole 16E5
Maych'ew 34D2
Mayenne 18D2
Maykop 23I7
Mayna 23J5
Mayotte terr. 35E5
Mayskiy 23J8
Mazara del Vallo 20E6
Mazār-e Sharīf 26F3
Mazatlán 50C4
Mažeikiai 11M8
Mazyr 23F5
Mbabane 37J4
Mbahiakro 32C4
Mbaïki 34B3
Mbale 34D3
Mbalmayo 32E4
Mbandaka 34B4
M'banza Congo 35B4
Mbarara 33G5
Mbari r. 34C3
Mbeya 35D4
Mbinga 35D5
Mbombela 37J3
Mbouda 32E4
Mbour 32B3
Mbuji-Mayi 35C4
Mburucuyá 54E3
McAlester 47H5
McAllen 46H6
McCall 46D3
McCook 46G3
Mchinga 35D4
McKinley, Mount 44C3
McMinnville 46C2
McPherson 46H4
Mdantsane 37H7
M'Daourouch 20B6
Mead, Lake resr 49E2
Meadville 48A2
Meaux 18F2
Mecca 34D1
Mechanicsville 48C4
Mechelen 12J5
Mecheria 32C1
Mecklenburger Bucht b. 13M3
Medan 27I6
Médéa 19H5
Medellín 52C2
Medenine 32E1
Medford 48E2
Medford 46C3
Medgidia 21M2
Media 48D3
Mediaş 21K1
Medicine Bow Mountains 46F3
Medicine Hat 44G4
Medina 55C2
Medina 34D1
Medina de Rioseco 19D3
Mediterranean Sea 20D6
Medvedevo 22J4
Medvezh'yegorsk 22G3
Meekatharra 40D5
Meerut 27G4
Megalopoli 21J6
Mehrestān 26F3
Meiganga 33E4
Meihekou 30B4
Meiktila 27I4
Meiningen 13M5
Meißen 13N5
Meixi 30C3
Meizhou 28D5
Mek'elē 34D2
Mékhé 32B3
Meknès 32C1
Mekong r. 27J4
Mekong r. 29C6
Melaka 29C7
Melbourne 42B6
Melbourne 47K6
Melenki 23I5
Melilla 19E6
Melitopol' 23G7
Melk 13O6
Melksham 15E7
Melo 54F4
Melrose 16G5
Melton 42B6
Melton Mowbray 15G6
Melun 18F2
Melville Bay 41H2
Melville Island 40G2
Melville Island 45H2
Memmingen 13M7
Memphis 47I4
Mena 23G6
Mena 47I5
Mende 18F4
Mendefera 33G3
Mendeleyevsk 22L5
Mendi 38E2
Mendip Hills 15E7
Mendoza 54C4
Menemen 21L5
Menongue 35B5
Mentawai, Kepulauan is 29B8
Menton 18H5
Menzel Bourguiba 20C6
Menzel Temime 20D6
Meppen 13K4
Meqheleng 37H5
Merano 20D1
Merauke 29G8
Merced 49B2
Mercedes 54E3
Mercedes 54E4
Mercês 55C3
Mere 15E7
Merefa 23H6
Mergui Archipelago is 27I5
Mérida 50G4
Mérida 19C4
Mérida 52D2
Meriden 48E2
Meridian 47J5
Mérignac 18D4
Merimbula 42D6
Merowe 33G3
Merredin 40D6
Merrill 47J2
Mersa Fatma 33H3
Mersey est. 14E5
Mersin 33G1
Merthyr Tydfil 15D7
Merzig 13K6
Mesa 46E5
Mesagne 20G4
Mesolongi 21I5
Mesquita 55C2
Messina 20F5
Messina, Strait of strait 20F5
Messini 21J6
Mestre 20E2
Meta r. 52E2
Metán 54C3
Methuen 48F1
Metlaoui 32D1
Metu 34D3
Metz 18H2
Meuse r. 12J5
Mevagissey 15C8
Mexicali 49E4
Mexico country 50D4
Mexico, Gulf of 47H6
Mexico City 50E5
Meyersdale 48B3
Mezdra 21J3
Mezen' 22J2
Mezen' r. 22J2
Mezhdurechensk 24J4
Mezhdurechensk 22K3
Meztúr 21I1
Miami 47K6
Miami 47I4
Miami Beach 47K6
Mīāndoāb 33H1
Miandrivazo 35E5
Mianwali 27G3
Mianyang 27J3
Miarinarivo 35E5
Miass 24H4
Michalovce 23D6
Michigan state 47J2
Michigan, Lake 47J3
Michurinsk 23I5
Micronesia, Federated States of country 29G7
Middelburg 12I5
Middelburg 37I3
Middelfart 11F9
Middle River 48C3
Middlesbrough 14F4
Middletown 48E2
Middletown 48D2
Midland 47L3
Midland 46G5
Midleton 17D6
Miðvágur 10□1
Mielec 23D6
Miercurea Ciuc 21K1
Mieres del Camín 19D2
Miguel Auza 46G7
Mihara 31D6
Mikhaylov 23H5
Mikhaylovka 30D4
Mikhaylovka 23I6
Mikhaylovsk 23I7
Mikhaylovskoye 24I4
Mikkeli 11O6
Milan 20C2
Milas 21L6
Milazzo 20F5
Mildenhall 15H6
Mildura 41I6
Miles 42E1
Miles City 46F2
Milford 48F1
Milford Haven 15B7
Milford Sound inlet 43A7
Miliana 19H5
Mil'kovo 25Q4
Millau 18F4
Milledgeville 47K5
Mille Lacs, Lac des l. 45I5
Millerovo 23I6
Millmerran 42E1
Millville 48D3
Milpitas 49B2
Milton Keynes 15G6
Milwaukee 47J3
Mīnāb 26E4
Minahasa, Semenanjung pen. 29E7
Minas 27J6
Minas 54E4
Minas Gerais state 55B2
Minas Novas 55C2
Minatitlán 50F5
Mindanao i. 29E7
Mindelo 32□
Minden 13L4
Minden 47I5
Mindoro i. 29E6
Mindouli 34B4
Minehead 15D7
Mineola 48E2
Mineral'nyye Vody 23I7
Mineral Wells 46H5
Minerva 48A2
Mingäçevir 23J8
Minglanilla 19F4
Mingoyo 35D5
Mingshui 30B3
Minna 32D4
Minneapolis 47I3
Minnesota state 47I2
Minorca i. 19H3
Minot 46G2
Minsk 11O10
Mińsk Mazowiecki 13R4
Minusinsk 24K4
Mirabela 55B2
Miraí 55C3
Miramar 54E5
Miramichi 45L5
Miranda 54E2
Miranda de Ebro 19E2
Mirandela 19C3
Mirandola 20D2
Mirassol 55A3
Mirboo North 42C7
Miri 29D7
Mirim, Lagoa l. 54F4
Mirnyy 25M3
Mirpur Khas 27F4
Miryang 31C6
Mirzapur 27H4
Miskolc 23D6
Mişrātah 33E1
Mission Viejo 49D4
Mississauga 48B1
Mississippi r. 47J6
Mississippi state 47J5
Missoula 46E2
Missouri r. 47I4
Missouri state 47I4
Mistassini, Lac l. 45K4
Mistelbach 13P6
Mitchell 41J5
Mitchell r. 41I3
Mitchell 46H3
Mitchelstown 17D5
Mito 31F5
Mitrovicë 21I3
Mitú 52D3
Mitumba, Chaîne des mts 35C5
Miura 31E6
Miyako 31F5
Miyakonojō 31C7
Miyazaki 31C7

Miyazu 31D6
Miyoshi 31D6
Mizen Head 17C6
Mizhhir"ya 23D6
Mjölby 11I7
Mkata 35D4
Mladá Boleslav 13O5
Mladenovac 21I2
Mława 13R4
Mlungisi 37H6
Mmabatho 37G3
Moanda 34B4
Moberly 47I4
Mobile 47J5
Mobile Bay 47J5
Moçambique 35E5
Mocha 34E2
Mochudi 37H3
Mocimboa da Praia 35E5
Mocoa 52C3
Mococa 55B3
Mocuba 35D5
Modder r. 37G5
Modena 20D2
Modesto 49B2
Moe 42C7
Moelv 11G6
Moffat 16F5
Mogadishu 34E3
Mogi das Cruzes 55B3
Mogi Mirim 55B3
Mogocha 25M4
Mogoditshane 37G3
Mohács 20H2
Mohale's Hoek 37H6
Mohammadia 19G6
Mohoro 35D4
Mohyliv-Podil's'kyy 23E6
Moinești 21L1
Mo i Rana 10I3
Mojave Desert 49D3
Mokhotlong 37I5
Moknine 20D7
Mokolo 33E3
Mokopane 37I3
Mokpo 31B6
Mokrous 23J6
Mokshan 23J5
Molde 10E5
Moldova country 23F7
Moldovei de Sud, Cîmpia plain 21M1
Molepolole 37G3
Molėtai 11N9
Molfetta 20G4
Molina de Aragón 19F3
Mollendo 52D7
Mölnlycke 11H8
Molong 42D4
Molopo watercourse 36E5
Moloundou 33E4
Moluccas is 29E8
Mombaça 53K5
Mombasa 34D4
Momchilgrad 21K4
Mompós 52D2
Møn i. 11H9
Monaco country 18H5
Monadhliath Mountains 16E3
Monaghan 17F3
Monastir 20D7
Monastyryshche 23F6
Monbetsu 30F3
Moncalieri 20B2
Monchegorsk 10R3
Mönchengladbach 13K5
Monclova 46G6
Moncton 45L5
Mondovì 20B2
Mondragone 20E4
Monfalcone 20E2
Monforte de Lemos 19C2
Mongbwalu 34D3
Móng Cai 27J4
Mongo 33E3
Mongolia country 27J2
Mongu 35C5
Monkey Bay 35D5
Monmouth 15E7
Monopoli 20G4
Monreal del Campo 19F3
Monreale 20E5
Monroe 47I5
Monrovia 32B4
Mons 12I5
Montana 21J3
Montana state 46F2
Montargis 18F3
Montauban 18E4
Montbrison 18G4
Montceau-les-Mines 18G3
Mont-de-Marsan 18D5
Monte Alegre 53H4
Monte Alegre de Goiás 55B1
Monte Alegre de Minas 55A2
Monte Azul 55C1
Monte Azul Paulista 55A3
Montebelluna 20E2
Monte-Carlo 18H5
Monte Cristi 51J5
Montego Bay 51I5
Montélimar 18G4
Montemorelos 46H6
Montemor-o-Novo 19B4
Montenegro country 20H3
Monterey 49B2
Monterey Bay 49A2
Montería 52C2
Monteros 54C3
Monterrey 46G6
Monte Santo 53K6
Montes Claros 55C2
Montesilvano 20F3
Montevarchi 20D3
Montevideo 54E4
Montgomery 47J5
Montgomery 48A3
Monthey 18H3
Monticello 48D2
Montilla 19D5
Montluçon 18F3
Montmagny 45K5
Monto 41K4
Montpelier 47M3
Montpellier 18F5
Montréal 45K5
Montrose 16G4
Montrose 46F4
Mont-St-Aignan 15I9
Montserrat terr. 51L5
Monywa 27I4
Monza 20C2
Moora 40D6
Moorhead 47H2
Mooroopna 42B6
Moose Jaw 44H4
Mopipi 35C6
Mopti 32C3
Moquegua 52D7
Mora 33E3
Mora 11I6
Morada Nova 53K5
Moramanga 35E5
Morar, Loch l. 16D4
Moray Firth b. 16E3
Mordovo 23I5
Morecambe 14E4
Morecambe Bay 14D4
Moree 42D2
Morelia 50D5
Morella 19F3
Morena, Sierra mts 19C5
Moreni 21K2
Moreno Valley 49D4
Morgan Hill 49B2
Morganton 47K4
Morgantown 48B3
Morges 18H3
Morioka 31F5
Morisset 42E4
Morki 22K4
Morlaix 18C2
Morley 14F5
Mornington Island 41H3
Morocco country 32C1
Morogoro 35D4
Moro Gulf 29E7
Morombe 35E6
Mörön 27J2
Morondava 35E6
Morón de la Frontera 19D5
Moroni 35E5
Moroto 34D3
Morozovsk 23I6
Morpeth 14F3
Morrinhos 55A2
Morristown 48D2
Morristown 47K4
Morrisville 48D1
Morro do Chapéu 53J6
Morshansk 23I5
Morsott 20C7
Morteros 54D4
Mortlake 42A7
Moruya 42E5
Morvern reg. 16D4
Morwell 42C7
Mosbach 13L6
Moscow 22H5
Moscow 46D2
Moselle r. 18H2
Moses Lake 46D2
Moshi 34D4
Mosjøen 10H4
Mosonmagyaróvár 13P7
Mosquitos, Golfo de los b. 51H7
Moss 11G7
Mossel Bay 36F8
Mossman 41J3
Mossoró 53K5
Most 13N5
Mostaganem 19G6
Mostar 20G3
Mostovskoy 23I7
Mosul 33H1
Motala 11I7
Motherwell 16F5
Motilla del Palancar 19F4
Motril 19E5
Motru 21J2
Mottama, Gulf of 27I5
Motul 50G4
Mouila 34B4
Moulins 18F3
Moultrie 47K5
Moundou 33E4
Moundsville 48A3
Mountain Home 47I4
Mountain Home 46D3
Mount Darwin 35D5
Mount Gambier 41I7
Mount Hagen 38E2
Mount Holly 48D3
Mount Isa 41H4
Mount Magnet 40D5
Mountmellick 17E4
Mount Morris 48C1
Mount Pleasant 47I3
Mount Pleasant 47K3
Mount Pleasant 47I5
Mount's Bay 15B8
Mount Shasta 46C3
Mount Vernon 47J4
Mount Vernon 46C2
Moura 41J4
Moura 19C4
Mourdi, Dépression du depr. 33F3
Mourdiah 32C3
Mourne Mountains hills 17F3
Mouscron 12I5
Mouydir, Monts du plat. 32D2
Moyeni 37H6
Moyobamba 52C5
Mozambique country 35D6
Mozambique Channel strait 35E6
Mozdok 23J8
Mozhaysk 23H5
Mozhga 22L4
Mpanda 35D4
Mpika 35D5
Mpumalanga prov. 37I4
Mpwapwa 35D4
M'Saken 20D7
M'Sila 19I6
Mstsislaw 23F5
Mthatha 37I6
Mtsensk 23H5
Mtwara 35E5
Mubende 34D3
Mubi 32E3
Muconda 35C5
Mucuri 55D2
Mudanjiang 30C3
Mudanya 21M4
Mudurnu 21N4
Mufulira 35C5
Muğla 21M6
Mühlhausen/Thüringen 13M5
Muir of Ord 16E3
Muju 31B5
Mukacheve 23D6
Mukalla 34E2
Mulan 30C3
Mulhacén mt. 19E5
Mulhouse 18H3
Muling 30C3
Mull i. 16D4
Mull, Sound of sea chan. 16C4
Mullens 48A4
Mullingar 17E4
Mull of Kintyre hd 16D5
Mullumbimby 42F2
Multan 27G3
Mumbai 27G5
Muna 50G4
Mundesley 15I6
Mundubbera 41K5
Munger 27H4
Munich 13M6
Munster 13M4
Münster 13K5
Muqui 55C3
Murakami 31E5
Muramvya 34C4
Murashi 22K4
Muratlı 21L4
Murcia 19F5
Murcia aut. comm. 19F5
Mureșul r. 21I1
Muret 18E5
Murfreesboro 47J4
Muriaé 55C3
Müritz l. 13N4
Murmansk 10R2
Murmanskaya Oblast' admin. div. 10S2
Murom 22I5
Muroran 30F4
Muroto 31D6
Murray r. 42A5
Murray 47J4
Murray Bridge 41H7
Murrumbidgee r. 42A5
Murrumburrah 42D5
Murska Sobota 20G1
Murzūq 33E2
Mürzzuschlag 13O7
Musala mt. 21J3
Musan 30C4
Muscat 26E4
Muscatine 47I3
Musgrave Ranges mts 40G5
Mushie 34B4
Muskogee 47H4
Musoma 34D4
Musselburgh 16F5
Mustafakemalpaşa 21M4
Muswellbrook 42E4
Mūţ 33F2
Mutare 35D5
Mutoko 35D5
Mutsamudu 35E5
Mutsu 30F4
Mutum 55C2
Muyinga 34D4
Muzaffarpur 27H4
Múzquiz 46G6
Mvuma 35D5
Mwanza 34D4
Mwene-Ditu 35C4
Mweru, Lake 35C4
Myadzyel 11O9
Myanmar country 27I4
Myeik 27I5
Myingyan 27I4
Myitkyina 27I4
Mykolayiv 21O1
Mykonos i. 21K6
Mymensingh 27I4
Myŏnggan 30C4
Myory 11O9
Mýrdalsjökull ice cap 10□2
Myrhorod 23G6
Myronivka 23F6
Myrtle Beach 47L5
Myrtleford 42C6
Myrtoo Pelagos sea 21J6
Myślibórz 13O4
Mysuru 27G5
My Tho 29C6
Mytilini 21L5
Mytilini Strait strait 21L5
Mytishchi 22H5
Mzimba 35D5
Mzuzu 35D5

N

Naantali 11M6
Naas 17F4
Nabari 31E6
Naberera 35D4
Naberezhnyye Chelny 24G4
Nabeul 20D6
Nacala 35E5
Nachingwea 35D5
Nacogdoches 47I5
Nador 19E6
Nadvirna 23E6
Nadym 24I3
Næstved 11G9
Nafplio 21J6
Naftalan 23J8
Nafy 34E1
Naga 29E6
Nagambie 42B6
Nagano 31E5
Nagaoka 31E5
Nagasaki 31C6
Nagato 31C6
Nagercoil 27G6
Nagoya 31E6
Nagpur 27G4
Nagqu 27I3
Nagyatád 20G1
Nagykanizsa 20G1
Nahāvand 33H1
Nairn 16F3
Nairobi 34D4
Naji 30A2
Najin 30C4
Najrān 34E2
Nakasongola 33G4
Nakatsu 31C6
Nakatsugawa 31E6
Nakfa 33G3
Nakhodka 30D4
Nakhon Pathom 29C6
Nakhon Ratchasima 29C6
Nakhon Sawan 29C6
Nakhon Si Thammarat 29B7
Nakskov 11G9
Nakuru 34D4
Nal'chik 23I8
Nallıhan 21N4
Nālūt 32E1
Namahadi 37I4
Namangan 27G2
Nambour 42F1
Nambucca Heads 42F3
Nam Định 29C5
Namib Desert 36B3
Namibe 35B5
Namibia country 35B6
Nampa 46D3
Nampala 32C3
Nampo 31B5
Nampula 35D5
Namsos 10G4
Namtsy 25N3
Namtu 27I4
Namur 12J5
Namwon 31B6
Nan 29C6
Nanaimo 44F5
Nanao 31E5
Nancha 30C3
Nanchang 27K4
Nanchong 27J3
Nancy 18H2
Nanded 27G5
Nanga Eboko 32E4
Nangulangwa 35D4
Nanjing 28D4
Nan Ling mts 27K4
Nanning 27J4
Nanortalik 45N3
Nanping 28D5
Nantes 18D3
Nanticoke 48A1
Nantucket Sound g. 48F2
Nantwich 15E5
Nanuque 55C2
Nanyang 27K3
Nanyuki 34D4
Napa 49A1
Napier 43F4
Naples 20F4
Naples 47K6
Nara 31D6
Nara 32C3
Narach 11O9
Naracoorte 41I7
Naranjal 52C4
Narbonne 18F5
Nardò 20H4
Nares Strait strait 45K2
Narimanov 23J7
Narmada r. 27G4
Narni 20E3
Narodnaya, Gora mt. 24G3
Naro-Fominsk 23H5
Narooma 42E6
Narowlya 23F6
Närpes 10L5
Narrandera 42C5
Narromine 42D4
Narsaq 45N3
Nartkala 23I8
Naruto 31D6
Narva 11P7
Narva Bay 11O7
Narvik 10J2
Nar'yan-Mar 22L2
Naryn 27G2
Nashik 27G4
Nashua 48F1
Nashville 47J4
Nassau 47L6
Nasser, Lake resr 33G2
Nässjö 11I8
Nasushiobara 31F5
Nata 35C6
Natal 53K5
Natchez 47I5
Natchitoches 47I5
National City 49D4
Natitingou 32D3
Natividade 53I6
Natori 31F5
Natron, Lake salt l. 34D4
Natuna Besar i. 27J6
Naturaliste, Cape 40D6
Naturaliste Channel 40C5
Naujoji Akmen 11M8
Naurskaya 23J8
Nauru country 39G2
Navahrudak 11N10
Navan 17F4
Navapolatsk 11P9
Navarra aut. comm. 19F2
Navashino 22I5
Navassa Island terr. 51I5
Navlya 23G5
Năvodari 21M2
Navoiy 26F2
Navojoa 46F6
Navolato 46F7
Nawabshah 26F4
Naxçıvan 26D3
Naxos i. 21K6
Nayoro 30F3
Nazaré 55D1
Nazário 55A2
Nazca 52D6
Nazerat 33G1
Nazilli 21M6
Nazran' 23J8
Nazrēt 34D3
Nazwá 26E4
N'dalatando 35B4
Ndélé 34C3
Ndjamena 33E3
Ndola 35C5
Neagh, Lough l. 17F3
Nea Liosia 21J5
Neapoli 21J6
Neath 15D7
Nebbi 34D3
Nebolchi 22G4
Nebraska state 46G3
Nebraska City 47H3
Necochea 54E5
Nédroma 19F6
Neftegorsk 23K5
Neftegorsk (abandoned) 30F1
Neftekamsk 24G4
Nefteyugansk 24I3
Negage 35B4
Negēlē 34D3
Negotino 21J4
Negro r. 55A4
Negro r. 52G4
Negros i. 29E7
Nehe 30B2
Neijiang 27J4
Nei Mongol Zizhiqu aut. reg. 30A2
Neiva 52C3
Nek'emtē 34D3
Nekrasovskoye 22I4
Nelidovo 22G4
Nellore 27G5
Nelson 44G5
Nelson 43D5
Nelson 14E5
Néma 32C3
Neman 11M9
Nemours 18F2
Nemuro 30G4
Nemuro-kaikyō sea chan. 30G4
Nemyriv 23F6
Nenagh 17D5
Nenjiang 30B2
Neosho 47I4
Nepal country 27H4
Nerang 42F1
Nerchinsk 25M4
Nerekhta 22I4
Nerópolis 55A2
Neryungri 25N4
Ness, Loch l. 16E3
Nestos r. 21K4
Netherlands country 12J4
Neubrandenburg 13N4
Neuchâtel 18H3
Neuchâtel, Lac de l. 18H3
Neufchâtel-Hardelot 15I8
Neumünster 13L3
Neunkirchen 13P7
Neunkirchen 13K6
Neuquén 54C5
Neuruppin 13N4
Neusiedler See l. 13P7
Neustrelitz 13N4
Neuwied 13K5
Nevada 47I4
Nevada state 46D4
Nevada, Sierra mts 19E5
Nevada, Sierra mts 46C3
Nevel' 22F4
Nevel'sk 30F3
Never 30B1
Nevers 18F3
Nevesinje 20H3
Nevinnomyssk 23I7
Newala 35D5
New Albany 47J4
New Amsterdam 53G2
Newark 48D3
Newark 48D2
Newark-on-Trent 15G5
New Bedford 48F2
New Bern 47L4
Newberry 47K5
New Boston 47I5
New Braunfels 46H6
New Britain i. 38E2
New Britain 48E2
New Brunswick prov. 45L5
New Brunswick 48D2
Newburgh 48D2
Newbury 15F7
Newburyport 48F1
New Caledonia terr. 39G4
Newcastle 42E4
Newcastle 37I4
Newcastle 17G3
New Castle 48A2
Newcastle-under-Lyme 15E5
Newcastle upon Tyne 14F4
Newcastle West 17C5
New City 48E2
New Cumnock 16E5
New Delhi 27G4
New England Range mts 42E3
Newent 15E7
Newfane 48B1
Newfoundland i. 45M5
Newfoundland and Labrador prov. 45M4
New Guinea i. 38E2
New Halfa 33G3
New Hampshire state 48F1
Newhaven 15H8
New Haven 48E2
New Iberia 47I5
New Ireland i. 38F2
New Jersey state 48D3
New Kensington 48B2
New Liskeard 45K5
New London 48E2
Newman 40D4
Newmarket 15H6
New Martinsville 48A3
New Mexico state 46F5
New Orleans 47I6
New Philadelphia 48A2
New Plymouth 43E4
Newport 15E6
Newport 15F8
Newport 15D7
Newport 47I4
Newport 46C3
Newport 48F2
Newport Beach 49D4
Newport News 48C4
Newport Pagnell 15G6
Newquay 15B8
New Roads 47I5
New Rochelle 48E2
New Ross 17F5
Newry 17F3
New Siberia Islands 25P2
New South Wales state 42C4
New Stanton 48B2
Newton 47I3
Newton 47H4
Newton 48F1
Newton 48D2
Newton Abbot 15D8
Newton Stewart 16E6
Newtown 15D6
New Town 46G2
Newtownabbey 17G3
Newtownards 17G3
Newtown Mount Kennedy 17F4
New Ulm 47I3
New York 48E2
New York state 48D1
New Zealand country 43D5
Neya 22I4
Neyrīz 26E4
Neyshābūr 26E3
Ngaoundal 32E4
Ngaoundéré 33E4
Nguigmi 32E3
Nguru 32E3
Ngwelezana 37J5
Nha Trang 29C6
Nhill 41I7
Nhlangano 37J4
Niafounké 32C3
Niagara Falls 48B1
Niagara Falls 48B1
Niamey 32D3
Nianzishan 30A3
Nias i. 29B7
Nicaragua country 51G6
Nicaragua, Lake 51G6
Nicastro 20G5
Nice 18H5
Nice 49A1
Nicobar Islands 27I6
Nicosia 33G1
Nida 11L9
Nidzica 13R4
Nienburg (Weser) 13L4
Nieuw Nickerie 53G2
Niğde 26C3
Niger country 32D3
Niger r. 32D4
Nigeria country 32D4
Niigata 31E5
Niihama 31D6
Niimi 31D6
Niitsu 31E5
Nijmegen 13J5
Nikel' 10Q2
Nikkō 31E5
Nikolayevka 23J5
Nikolayevsk 23J6
Nikolayevsk-na-Amure 30F1
Nikol'sk 22J4
Nikopol' 23G7
Nikšić 20H3
Nile r. 33G1
Niles 48A2
Nîmes 18G5
Nimmitabel 42D6
Ning'an 30C3
Ningbo 28E5
Ninohe 31F4
Niono 32C3
Nioro 32C3
Niort 18D3
Nipigon, Lake 45J5
Nipissing, Lake 45K5
Niquelândia 55A1
Nirmal 27G5
Niš 21I3
Niscemi 20F6
Niterói 55C3
Nitra 13Q6
Niue terr. 39J3
Nivala 10N5
Nizamabad 27G5
Nizhnekamsk 22K5
Nizhneudinsk 25K4
Nizhnevartovsk 24I3
Nizhneyansk 25O2
Nizhniy Lomov 23I5
Nizhniy Novgorod 22I4
Nizhnyaya Omra 22L3
Nizhnyaya Tunguska r. 24J3
Nizhyn 23F6
Njombe 35D4
Njurundabommen 10J5
Nkambe 32E4
Nkawkaw 32C4
Nkhata Bay 35D5
Nkhotakota 35D5
Nkongsamba 32D4
Nobeoka 31C6
Noboribetsu 30F4
Nogales 46E5
Nogales 46E5
Nōgata 31C6
Nogent-le-Rotrou 18E2
Noginsk 22H5
Nogliki 30F2
Nola 34B3
Nolinsk 22K4
Nome 44B3
Nong'an 30B3
Noranda 45K5
Norberg 11I6
Norden 13K4
Norderstedt 13M4
Nordhausen 13M5
Norfolk 47H3
Norfolk 48C4
Norfolk Island terr. 39G4
Noril'sk 24J3
Norman 47H4
Normandia 53G3
Normandy reg. 18D2
Normanton 41I3
Norra Kvarken strait 10L5
Norristown 48D2
Norrköping 11J7
Norrtälje 11K7
Nortelândia 53G6
Northallerton 14F4
Northampton 15G6
Northampton 48E1
North Bay 45K5
North Canton 48A2
North Cape 10N1
North Cape 43D2
North Carolina state 47L4
North Channel lake channel 47K2
North Channel strait 17G2
North Dakota state 46G2
North Downs hills 15G7
Northern Cape prov. 36D5
Northern Donets r. 23I7
Northern Ireland prov. 17F3
Northern Mariana Islands terr. 29G6
Northern Territory admin. div. 40G3
North Frisian Islands 13L3
North Haven 48E2
North Island 43D4
North Kingsville 48A2
North Korea country 31B5
North Las Vegas 49E2
North Macedonia country 21I4
North Platte 46G3
North Platte r. 46G3
North Ronaldsay Firth sea chan. 16G1
North Saskatchewan r. 44H4
North Sea 12H2
North Shields 14F3
North Taranaki Bight b. 43E4
North Tonawanda 48B1
North Tyne r. 14E4
North Uist i. 16B3
North West prov. 36G4
Northwest Territories admin. div. 44H3
Northwich 14E5
North York Moors moorland 14G4
Norton Sound sea chan. 44B3
Norwalk 48E2
Norwalk 48A2
Norway country 10E6
Norwegian Sea 10E3
Norwich 15I6
Norwich 48E2
Norwich 48D1
Noshiro 31F4
Nossob watercourse 36D2
Notodden 11F7
Notre-Dame, Monts mts 45L5
Notre Dame Bay 45M5
Nottingham 15F6
Nouâdhibou 32B2
Nouakchott 32B3
Nouméa 39G4
Nouna 32C3
Nouvelle Calédonie i. 39G4
Nova Friburgo 55C3
Nova Iguaçu 55C3
Nova Kakhovka 21O1
Nova Lima 55C2
Nova Odesa 23F7
Nova Ponte 55B2
Novara 20C2
Nova Scotia prov. 45L5
Novato 49A1
Nova Trento 55A4
Nova Venécia 55C2
Nova Xavantina 53H6
Novaya Sibir', Ostrov i. 25P2
Novaya Zemlya is 24G2
Nova Zagora 21L3
Novelda 19F4
Nové Zámky 13Q7
Novhorod-Sivers'kyy 23G6
Novi Ligure 20C2
Novi Pazar 21L3
Novi Pazar 21I3
Novi Sad 21H2
Novoanninskiy 23I6
Novo Aripuanã 52F5
Novoazovs'k 23H7
Novocheboksarsk 22J4
Novocherkassk 23I7
Novo Cruzeiro 55C2
Novodvinsk 22I2
Novo Hamburgo 55A5
Novohrad-Volyns'kyy 23E6
Novokhopersk 23I6
Novokiyevskiy Uval 30C2
Novokubansk 23I7
Novokuybyshevsk 23K5
Novokuznetsk 24J4
Novo mesto 20F2
Novomoskovsk 23H5
Novomoskovs'k 23G6
Novonikolayevskiy 23I6
Novopokrovskaya 23I7
Novopskov 23H6
Novorossiysk 23H7
Novorzhev 22F4
Novoshakhtinsk 23H7
Novosibirsk 24J4
Novosokol'niki 22F4
Novospasskoye 23J5
Novotroyits'ke 23G7
Novoukrayinka 23F6
Novouzensk 23K6
Novovolyns'k 23E6
Novovoronezh 23H6
Novozybkov 23F5
Nový Jičín 13P6
Novyy Oskol 23H6
Novyy Urengoy 24I3
Nowra 42E5
Nowy Sącz 13R6
Nowy Targ 13R6
Noyabr'sk 24I3
Nsanje 35D5
Nsukka 32D4
Ntungamo 34D4
Nuba Mountains 33G3
Nubian Desert 33G2
Nueva Gerona 51H4
Nueva Imperial 54B5
Nueva Loja 52C3
Nueva Rosita 46G6
Nuevitas 51I4
Nuevo Casas Grandes 46F5
Nuevo Laredo 46H6
Nuku'alofa 39I4
Nukus 26E2
Nullarbor Plain 40F6
Numan 34B3
Numazu 31E6
Numurkah 42B6
Nunavut admin. div. 45I3
Nuneaton 15F6
Nunivak Island 44B4
Nuñomoral 19C3
Nuoro 20C4
Nuqrah 34E1
Nuremberg 13M6
Nurlat 23K5
Nurmes 10P5
Nurmo 10M5
Nuuk 45M3
Nuweveldberge mts 36E7
Nyagan' 24H3
Nyala 33F3
Nyamtumbo 35D5
Nyandoma 22I3
Nyasa, Lake 35D4
Nyasvizh 11O10
Nyborg 11G9
Nybro 11I8
Nyeri 34D4
Nyíregyháza 23D7
Nykøbing Falster 11G9
Nykøbing Sjælland 11G9
Nyköping 11J7
Nynäshamn 11J7
Nyngan 42C3
Nyon 18H3
Nysa 13P5
Nyunzu 35C4
Nyurba 25M3
Nzega 35D4
Nzérékoré 32C4

O

O'ahu i. 46□
Oakey 42E1
Oakham 15G6
Oakland 49A2
Oakridge 46C3
Oakville 48B1
Oamaru 43C7
Oaxaca 50E5
Ob' r. 24H3
Obama 31D6
Oban 16D4
Oberon 42D4
Oberpfälzer Wald mts 13N6
Óbidos 53G4
Obihiro 30F4
Obil'noye 23J7
Obluch'ye 30C2
Obninsk 23H5
Obo 34C3
Obock 34E2
Oboyan' 23H6
Obrenovac 21I2
Obskaya Guba sea chan. 24I3
Obuasi 32C4
Ob"yachevo 22K3
Ocala 47K6
Ocaña 52D2
Ocaña 19E4
Occidental, Cordillera mts 52C3
Occidental, Cordillera mts 52D7
Ocean City 48D3
Oceanside 49D4
Ochakiv 21N1
Ochamchire 23I8
Ochil Hills 16F4
Octeville-sur-Mer 15H9
Ódáðahraun lava field 10□2
Ōdate 31F4
Odawara 31E6
Odda 11E6
Ödemiş 21L5
Odense 11G9
Oderbucht b. 13O3
Odessa 21N1
Odessa 46G5
Odienné 32C4
Odintsovo 22H5
Odra r. 13Q6
Oeiras 53J5
Of 23I8
Offa 32D4
Offenbach am Main 13L5
Offenburg 13K6
Oga 31E5
Ōgaki 31E6
Ogallala 46G3
Ogbomoso 32D4
Ogden 46E3
Ogre 11N8
Ohio r. 48A3
Ohio state 48A2
Ohrid 21I4
Ohrid, Lake 21I4
Oil City 48B2
Ōita 31C6
Ojinaga 46G6
Ojiya 31E5
Ojos del Salado, Nevado mt. 54C3
Oka r. 23I4
Okavango Delta swamp 35C5
Okaya 31E5
Okayama 31D6
Okazaki 31E6
Okeechobee, Lake 47K6
Okehampton 15C8
Okha 30F1
Okhotsk 25P4
Okhotsk, Sea of 30G3
Okhtyrka 23G6
Okinawa-shotō is 31B8
Oklahoma state 46H4
Oklahoma City 47H4
Okmulgee 47H4
Okovskiy Les for. 22G5
Oktyabr'skiy 22I3
Oktyabr'skiy 25Q4
Oktyabr'skiy 24G4
Oktyabr'skiy 23I7
Oktyabr'skoy Revolyutsii, Ostrov i. 25K2
Okulovka 22G4
Okuta 32D4
Öland i. 11J8
Olavarría 54D5
Oława 13P5
Olbia 20C4
Oldenburg 13L4
Oldham 14E5
Old Head of Kinsale 17D6
Olean 48B1
Olecko 13S3
Olekminsk 25N3
Oleksandriya 23G6
Olenegorsk 10R2
Olenino 22G4
Olevs'k 23E6
Olhão 19C5
Olifants r. 37J3
Olifants r. 36D6
Olifants r. 36E7
Olinda 53L5
Oliva 19F4
Oliva, Cordillera de mts 54C3
Oliveira dos Brejinhos 55C1
Olmos 52C5
Olney 15G6
Olney 48C3
Olofström 11I8
Olomouc 13P6
Olonets 22G3
Oloron-Ste-Marie 18D5
Olot 19H2
Olovyannaya 25M4
Olsztyn 13R4
Olten 18H3
Oltenița 21L2
Olympia 46C2
Olympus, Mount 21J4
Omagh 17E3
Omaha 47H3
Omaheke admin. reg. 36D2
Oman country 26E5
Oman, Gulf of 26E4
Omdurman 33G3
Om Hajër 33G3
Ōmiya 31E6
Omsk 24I4
Omsukchan 25Q3
Ōmura 31C6
Omutninsk 22L4
Oncativo 54D4
Ondjiva 35B5
Ondo 32D4
Öndörhaan 27K2
Onega 22H3
Onega r. 22H3
Onega, Lake 22G3
Oneida 48D1
Oneonta 48D1
Onești 21L1
Onezhskaya Guba g. 22G2
Ongjin 31B5
Ongole 27H5
Onitsha 32D4
Onomichi 31D6
Ontario prov. 48A1
Ontario 49D3
Ontario, Lake 48C1
Onverwacht 53G2
Oparino 22K4
Opava 13P6
Opelika 47J5
Opelousas 47I5
Opochka 11P8
Opole 13P5
Oporto 19B3
Opuwo 35B5
Oradea 21I1
Oran 19F6
Orán 54D2
Orang 30C4
Orange 42D4
Orange 18G4
Orange r. 36C5
Orange 49D4
Orange 47I5
Orangeburg 47K5
Orange Walk 50G5
Oranjemund 36C5
Oranjestad 51J6
Orăștie 21J2
Orbost 42D6
Ordu 23H8
Ore 32D4
Örebro 11I7
Oregon state 46C3
Oregon City 46C2
Orekhovo-Zuyevo 22H5
Orel 23H5
Orenburg 24G4
Øresund strait 11H9
Orford Ness hd 15I6
Orhaneli 21M5
Orhangazi 21M4
Orichi 22K4
Oriental, Cordillera mts 52D2
Oriental, Cordillera mts 52E6
Orihuela 19F4
Orikhiv 23G7
Orimattila 11N6
Orinoco r. 52F2
Oristano 20C5
Orivesi 11N6
Oriximiná 53G4
Orizaba 50E5
Orizona 55A2
Orkney 37H4
Orkney Islands 16F1
Orlândia 55B3
Orlando 47K6
Orleans 55A5
Orléans 18E3
Orlov 22K4
Orlovskiy 23I7
Ormskirk 14E5
Örnsköldsvik 10K5
Orodara 32C3
Oroquieta 29E7
Orosháza 21I1
Oroville 46C4
Orqohan 30A2
Orsha 23F5
Orshanka 22J4
Orsk 24G4
Ørsta 10E5
Orthez 18D5
Ortona 20F3
Oruro 52E7
Orvieto 20E3
Ōsaka 31D6
Ōsaki 31F5
Osby 11H8
Osh 27G2
Oshakati 35B5
Oshawa 45K5
Oshkosh 47J3
Ōshū 31F5
Osijek 20H2
Osimo 20E3
oSizweni 37J4
Oskarshamn 11J8
Oslo 11G7
Oslofjorden sea chan. 11G7
Osmancık 23G8
Osmaneli 21M4
Osnabrück 13L4
Osogbo 32D4
Osorno 54B6
Osorno 19D2
Osoyoos 46D2
Osøyro 11D6
Oss 13J5
Ossining 48E2
Ostashkov 22G4
Ostend 12I5
Östersund 10I5
Ostrava 13Q6
Ostróda 13Q4
Ostrogozhsk 23H6
Ostrov 11P8
Ostrovskoye 22I4
Ostrowiec Świętokrzyski 23D6
Ostrów Mazowiecka 13R4
Ostrów Wielkopolski 13P5
Oswego 48C1
Oswestry 15D6
Otago Peninsula 43C7
Otaki 43E5
Otaru 30F4
Ōtawara 31F5
Otepää 11O7
oThongathi 37J5
Otjiwarongo 35B6
Otjozondjupa admin. reg. 36C1
Otley 14F5
Otradnyy 23K5
Otranto, Strait of strait 20H4
Ōtsu 31D6
Ottawa 45K5
Ottawa 47J3
Ottawa 47H4
Ottumwa 47I3
Otway, Cape 42A7
Ouachita Mountains 47I5
Ouaddaï reg. 33F3
Ouagadougou 32C3
Ouahigouya 32C3
Oualâta, Ḍahr hills 32C3
Ouargaye 32D3
Ouargla 32D1
Ouarzazate 32C1
Oudtshoorn 36F7
Oued Zem 32C1
Oued Zénati 20B6
Ouesso 34B3
Ouezzane 19D6
Oujda 19F6
Oulainen 10N4
Ouled Djellal 19I6
Ouled Farès 19G5
Oulu 10N4
Oulujärvi l. 10O4
Oulunsalo 10N4
Oum el Bouaghi 20B7
Oundle 15G6
Ourense 19C2
Ouricuri 53J5
Ourinhos 55A3
Ouro Preto 55C3
Ouse r. 14G5
Ouse r. 15H8
Outapi 35B5
Outer Hebrides is 16B3
Outer Santa Barbara Channel 49C4
Outlook 46F1
Outokumpu 10P5
Ouyen 41I7
Ovalle 54B4
Ovar 19B3
Oviedo 19D2
Ovruch 23F6
Owando 34B4
Owase 31E6
Owatonna 47I3
Owego 48C1
Owensboro 47J4
Owen Sound 45J5
Owen Stanley Range mts 38E2
Owerri 32D4
Owo 32D4
Öxarfjörður b. 10□2
Oxelösund 11J7
Oxford 15F7
Oxford 47J5
Oxnard 49C3
Oyama 31E5
Oyem 34B3
Oyo 32D4
Oyonnax 18G3
Ozark 47J5
Ozark Plateau 47I4
Ozarks, Lake of the 47I4
Ozernovskiy 25Q4
Ozersk 11M9
Ozery 23H5
Ozinki 23K6

P

Paamiut 45N3
Paarl 36D7
Pabianice 13Q5
Pabna 27H4
Pacasmayo 52C5
Pachino 20F6
Pachuca 50E4
Pacific Grove 49B2
Padang 29C8
Padany 22G3
Paderborn 13L5
Padstow 15C8
Padua 20D2
Paducah 47J4
Pag 20F2
Pag i. 20F2
Pagadian 29E7
Paide 11N7
Paignton 15D8
Päijänne l. 11N6
Paimio 11M6
Painesville 48A2
Paisley 16E5
Paita 52B5
Pakaraima Mountains 52G3
Pakaraima Mountains 52F3
Pakhachi 25R3
Pakistan country 26F3
Pakruojis 11M9
Paks 20H1
Pakxé 29C6
Pala 33E4
Palaiseau 18F2
Palakkad 27G5
Palana 25Q4
Palanpur 27G4
Palapye 37H2
Palatka 25Q3
Palau country 29F7
Palawan i. 29D7
Palekh 22I4
Palembang 29C8
Palena 54B6
Palencia 19D2
Palermo 20E5
Palestine 47H5
Palhoça 55A4
Pali 27G4
Palikir 7
Palimé 32D4
Palk Strait strait 27G6
Pallasovka 23J6
Palliser, Cape 43E5
Palma de Mallorca 19H4
Palmares 53K5
Palmares do Sul 55A5
Palmas 55A4
Palmas 53I6
Palmas, Cape 32C4
Palmdale 49C3
Palmeira 55A4
Palmeira das Missões 54F3
Palmeira dos Índios 53K5
Palmeirais 53J5
Palmerston atoll 39J3
Palmerston North 43E5
Palmi 20F5
Palmira 52C3
Palm Springs 49D4
Palo Alto 49A2
Palopo 29E8
Palu 29D8
Pamiers 18E5
Pamir mts 27G3
Pamlico Sound sea chan. 47L4
Pampas reg. 54D5
Pamplona 52D2
Pamplona 19F2
Pamukova 21N4
Panagyurishte 21K3
Panaji 27G5
Panama country 51H7
Panamá, Canal de 51I7
Panama, Gulf of 51I7
Panama City 51I7
Panama City 47J5
Panay i. 29E6
Pančevo 21I2
Panevėžys 11N9
Pangkalanbuun 29D8
Pangkalpinang 29C8
Pangnirtung 45L3
Pankakoski 10Q5
Panorama 55A3
Panshi 30B4
Pantanal marsh 53G7
Pánuco 50E4
Panzhihua 27J4
Paoua 34B3
Pápa 20G1
Papakura 43E3
Papantla 50E4
Papa Stour i. 16□
Papenburg 13K4
Paphos 33G1
Papua, Gulf of 38E2
Papua New Guinea country 41J1
Par 15C8
Pará r. 55B2
Pará, Rio do r. 53I4
Paraburdoo 40D4
Paracatu 55B2
Paracel Islands 29D6
Paraćin 21I3
Paracuru 53K4
Pará de Minas 55B2
Paradise 46C4
Paragominas 53I4
Paragould 47I4
Paraguaçu Paulista 55A3
Paraguay r. 54E3
Paraguay country 54E2
Paraíso do Tocantins 53I6
Paraisópolis 55B3
Parakou 32D4
Paramaribo 53G2
Paramirim 55C1
Paramus 48D2
Paraná 54D4
Paranã 55B1
Paraná r. 55A1
Paraná state 55A4
Paraná r. 53H8
Paranaguá 55A4
Paranaíba 55A2
Paranapiacaba, Serra mts 55A4
Paranavaí 54F2
Paraopeba 55B2
Paratinga 55C1
Paraúna 55A2
Parbhani 27G5
Parchim 13M4
Pardubice 13O5
Parecis, Serra dos hills 52F6
Parepare 29D8
Pargas 11M6
Paria, Gulf of 51L6
Parintins 53G4
Paris 18F2
Paris 47H5
Parkano 11M5
Parkersburg 48A3
Parkes 42D4
Parma 20D2
Parma 48A2
Parnaíba 53J4
Pärnu 11N7
Paroikia 21K6
Paroo watercourse 42A3
Paros i. 21K6
Parral 54B5
Parramatta 42E4
Parras 46G6
Parry Channel 45G2
Parry Islands 45G2
Parsons 47H4
Parthenay 18D3
Partizansk 30D4
Pasadena 49C3
Pascagoula 47J5
Pașcani 21L1
Pashkovskiy 23H7
Paso de los Toros 54E4
Paso Robles 49B3
Passaic 48D2
Passa Tempo 55B3
Passau 13N6
Passo Fundo 54F3
Passos 55B3
Pastavy 11O9
Pasto 52C3
Pastos Bons 53J5
Pasvalys 11N8
Patagonia reg. 54B8
Paterson 48D2
Patna 27H4
Pato Branco 54F3
Patos 53K5
Patos, Lagoa dos l. 54F4
Patos de Minas 55B2
Patras 21I5
Patrocínio 55B2
Patton 48B2
Patu 53K5
Pau 18D5
Pauini 52E5
Paulo Afonso 53K5
Pavão 55C2
Pavia 20C2
Pavlikeni 21K3
Pavlodar 27G1
Pavlohrad 23G6
Pavlovka 23J5
Pavlovo 22I5
Pavlovsk 23I6
Pavlovskaya 23H7
Pawtucket 48F2
Payakumbuh 29C8
Payette 46D3
Paysandú 54E4
Pazar 23I8
Pazardzhik 21K3
Pazin 20E2
Peabody 48F1
Peace r. 44G4
Pearsall 46H6
Peary Channel 45I2
Peçanha 55C2
Pechenga 10Q2
Pechora 22M2
Pechora r. 22L1
Pechorskaya Guba b. 22L1
Pechory 11O8
Pecos 46G5
Pecos r. 46G6
Pécs 20H1
Pedernales 51J5
Pedra Azul 55C1
Pedregulho 55B3
Pedreiras 53J4
Pedro Juan Caballero 54E2
Peebles 16F5
Peekskill 48E2
Pegasus Bay 43D6
Pegu 27I5
Pehuajó 54D5
Peipus, Lake 11O7
Peixe 53I6
Peixoto de Azevedo 53H6
Pejë 21I3
Pekinga 32D3
Peloponnese admin. reg. 21J6
Pelotas 54F4
Pemba 35E5
Pembroke 45K5
Pembroke 15C7
Penápolis 55A3
Penarth 15D7
Penas, Golfo de g. 54A7
Pendleton 46D2
Penha 55A4
Peniche 19B4
Penicuik 16F5
Peninsular Malaysia 29C7
Penn Hills 48B2
Pennine, Alpi mts 18H4
Pennines hills 14E4
Pennsburg 48D2
Pennsville 48D3
Pennsylvania state 48B2
Penn Yan 48C1
Penonomé 51H7
Penrith 42E4
Penrith 14E4
Pensacola 47J5
Penticton 44G5
Pentland Firth sea chan. 16F2
Pentland Hills 16F5
Penza 23J5
Penzance 15B8
Peoria 47J3
Perdizes 55B2
Pereira 52C3
Pereira Barreto 55A3
Peremyshlyany 23E6
Pereslavl'-Zalesskiy 22H4
Pereyaslavka 30D3
Pereyaslav-Khmel'nyts'kyy 23F6
Pergamino 54D4
Perico 54C2
Périgueux 18E4
Perm' 24G4
Pernik 21J3
Perpignan 18F5
Perranporth 15B8
Perris 49D4
Perry 47K5
Perryton 46G4
Perryville 47J4
Perth 40D6
Perth 16F4
Perth Amboy 48D2
Peru country 52D6
Perugia 20E3
Pervomaysk 23I5
Pervomays'k 23F6
Pervomayskiy 23I5
Pervomays'kyy 23H6
Pesaro 20E3
Pescara 20F3
Peschanokopskoye 23I7
Peshawar 27G3
Peshkopi 21I4
Peshtera 21K3
Pesnica 20F1
Pessac 18D4
Pestovo 22G4
Pestravka 23K5
Petaluma 49A1
Petatlán 50D5
Peterborough 41H6
Peterborough 45K5
Peterborough 15G6
Peterborough 48F1
Peterhead 16H3
Peterlee 14F4
Petersburg 48C4
Petersfield 15G7
Petersville 44C3
Peto 50G4
Petoskey 47K2
Petrich 21J4
Petrolina 53J5
Petrolina de Goiás 55A2
Petropavlovsk-Kamchatskiy 25Q4
Petropavlovskoye 27F1
Petrópolis 55C3

Petroşani 21J2
Petrovsk 23J5
Petrovsk-Zabaykal'skiy 25L4
Petrozavodsk 22G3
Petukhovo 24H4
Petushki 22H5
Pevek 25S3
Pezinok 13P6
Pforzheim 13L6
Phahameng 37H5
Phalaborwa 37J2
Phangnga 29B7
Phan Rang-Thap Cham 29C6
Phan Thiêt 29C6
Phatthalung 29C7
Phenix City 47J5
Phet Buri 29B6
Philadelphia 48D3
Philippines *country* 29E6
Philippine Sea 29E6
Phiritona 37H4
Phitsanulok 29C6
Phnom Penh 29C6
Phoenix 46E5
Phoenix Islands 39I2
Phôngsali 28C5
Phônsavan 29C6
Phrae 29C6
Phuket 29B7
Piacenza 20C2
Piatra Neamț 21L1
Picardie *admin. reg.* 15J9
Picardy *reg.* 18E2
Picauville 15F9
Picayune 47J5
Pichanal 54D2
Pichilemu 54B4
Pickering 14G4
Picos 53J5
Pico Truncado 54C7
Picton 42E5
Piedade 55B3
Piedras Negras 46G6
Pieksämäki 10O5
Pielinen *l.* 10P5
Pierre 46G3
Pietermaritzburg 37J5
Pigg's Peak 37J3
Pihlajavesi *l.* 10P6
Pikalevo 22G4
Pikeville 47K4
Piła 13P4
Pilão Arcado 53J5
Pilar 54E4
Pilar 54E3
Pil'na 22J5
Pimenta Bueno 52F6
Pinamar 54E5
Pinar del Río 51H4
Pinarhisar 21L4
Piñas 52C4
Pińczów 13R5
Pindaí 55C1
Pindamonhangaba 55B3
Pindus Mountains 21I5
Pine Bluff 47I5
Pinega 22I2
Pinerolo 20B2
Pinetown 37J5
Pingdingshan 27K3
Pingxiang 27J4
Pingxiang 27K4
Pinheiro 53I4
Pinjarra 40D6
Pinsk 11O10
Pionki 13R5
Piotrków Trybunalski 13Q5
Piracanjuba 55A2
Piracicaba 55B3
Piracuruca 53J4
Piraeus 21J6
Piraí do Sul 55A4
Piraju 55A3
Pirajuí 55A3
Piranhas 53H7
Piranhas *r.* 53K5
Pirapora 55B2
Pirassununga 55B3
Pirenópolis 55A1
Pires do Rio 55A2
Piripiri 53J4
Pisa 20D3
Pisco 52C6
Písek 13O6
Pissis, Cerro 54C3
Pisté 50G4
Pistoia 20D3
Pita 32B3
Pitanga 55A4
Pitangui 55B2
Pitcairn Islands *terr.* 6
Piteå 10L4
Piterka 23J6
Pitești 21K2
Pitkyaranta 22F3
Pitlochry 16F4
Pittsburgh 48B2
Pittsfield 48E1
Pittsworth 42E1
Piumhi 55B3
Piura 52B5
Pivka 20F2
Pixley 49C3
Placerville 49B1
Plácido de Castro 52E6
Plainfield 48F2
Plainview 46G5
Planaltina 55B1
Planura 55A3
Plaquemine 47I5
Plasencia 19C3
Plato 52D2
Platte *r.* 46H3
Plattsburgh 47M3
Plauen 13N5
Plavsk 23H5
Playas 52B4
Pleasantville 48D3
Plenty, Bay of *g.* 43F3
Plesetsk 22I3
Pleven 21K3
Pljevlja 21H3
Płock 13Q4
Ploiești 21L2
Plovdiv 21K3
Plungė 11L9
Plymouth 15C8
Plymouth 47J3
Plymouth 48F2
Plymouth (abandoned) 51L5
Plynlimon *hill* 15D6
Plzeň 13N6
Pô 32C3
Po *r.* 20E2
Pocatello 46E3
Pochayiv 23E6
Pochep 23G5
Pochinki 23J5
Pochinok 23G5
Poções 55C1
Poconé 53G7
Poços de Caldas 55B3
Podgorenskiy 23H6
Podgorica 21H3
Podgornoye 24J4
Podol'sk 23H5
Podporozh'ye 22G3
Pofadder 36D5
Pogar 23G5
Poggibonsi 20D3
Pogradec 21I4
Pogranichnyy 30C3
Pohang 31C5
Pointe-à-Pitre 51L5
Pointe-Noire 35B4
Point Pleasant 48D2
Poitiers 18E3
Pokaran 27G4
Pokrovka 30C4
Pokrovsk 25N3
Pokrovskoye 23H7
Poland *country* 13Q4
Polatsk 11P9
Polessk 11L9
Police 13O4
Polis'ke (abandoned) 23F6
Polkowice 13P5
Polohy 23H7
Polokwane 37I2
Polonne 23E6
Poltava 23G6
Põlva 11O7
Polyarnyy 10R2
Polyarnyy (abandoned) 25S3
Polyarnyye Zori 10R3
Polygyros 21J4
Pombal 53K5
Pomeranian Bay 13O3
Pomezia 20E4
Pomona 49D3
Pomorie 21L3
Pompei 20F4
Pompéia 55A3
Ponazyrevo 22J4
Ponca City 47H4
Ponce 51K5
Ponferrada 19C2
Ponta Grossa 55A4
Pontal 55A3
Pontalina 55A2
Ponta Porã 54E2
Pontefract 14F5
Ponte Nova 55C3
Pontes e Lacerda 53G7
Pontevedra 19B2
Pontiac 47J3
Pontiac 47K3
Pontianak 29C8
Pontoise 18F2
Pontypool 15D7
Pontypridd 15D7
Poole 15F8
Poopó, Lago de *l.* 52E7
Popayán 52C3
Poplar Bluff 47I4
Popocatépetl, Volcán *vol.* 50E5
Popokabaka 35B4
Popovo 21L3
Poprad 13R6
Poquoson 48C4
Porangatu 55A1
Porbandar 27F4
Poreč 20E2
Porecatu 55A3
Poretskoye 23J5
Pori 11L6
Porirua 43E5
Porkhov 11P8
Porlamar 52F1
Poros 21J6
Porosozero 22G3
Porsangerfjorden *sea chan.* 10N1
Porsgrunn 11F7
Portadown 17F3
Portaferry 17G3
Portage 48B2
Portage 47J3
Portage la Prairie 45I5
Port Alberni 46C2
Portalegre 19C4
Portales 46G5
Port Angeles 46C2
Port Antonio 51I5
Portarlington 17E4
Port Arthur 47I6
Port Augusta 41H6
Port-au-Prince 51J5
Port Blair 27I5
Portbou 19H2
Port Chalmers 43C7
Port Charlotte 47K6
Port-de-Paix 51J5
Port Douglas 41J3
Porteirinha 55C1
Portel 53H4
Port Elizabeth 37G7
Porterville 49C2
Port-Gentil 34A4
Port Glasgow 16E5
Port Harcourt 32D4
Porthcawl 15D7
Port Hedland 40D4
Porthleven 15B8
Porthmadog 15C6
Port Hueneme 49C3
Port Huron 47K3
Portimão 19B5
Portland 42D4
Portland 48E1
Portland 47M3
Portland 46C2
Port-la-Nouvelle 18F5
Portlaoise 17E4
Port Lavaca 47H6
Port Lincoln 41H6
Port Loko 32B4
Port Louis 7
Port Macquarie 42F3
Port Moresby 41J1
Port Nolloth 36C5
Porto Alegre 55A5
Porto Amboim 35B5
Porto Belo 55A4
Porto de Moz 53H4
Porto Esperidião 53G7
Porto Franco 53I5
Port of Spain 51L6
Portogruaro 20E2
Portomaggiore 20D2
Porto Nacional 53I6
Porto-Novo 32D4
Porto Novo 32□
Porto Seguro 55D2
Porto Torres 20C4
Porto União 55A4
Porto-Vecchio 18I6
Porto Velho 52F5
Portoviejo 52B4
Port Phillip Bay 42B7
Port Pirie 41H6
Portree 16C3
Portrush 17F2
Port Said 33G1
Port St Joe 47J6
Port Shepstone 37J6
Portsmouth 15F8
Portsmouth 48F1
Portsmouth 47K4
Portsmouth 48C4
Port Sudan 33G3
Port Talbot 15D7
Portugal *country* 19C4
Port Vila 39G3
Porvenir 54B8
Porvoo 11N6
Posadas 54E3
Poshekhon'ye 22H4
Poso 29E8
Postmasburg 36F5
Potenza 20F4
Poti 23I8
Potiraguá 55D1
Potiskum 32E3
Potomac *r.* 48C3
Potosí 52E7
Potsdam 13N4
Potters Bar 15G7
Pottstown 48D2
Pottsville 48C2
Poughkeepsie 48E2
Poulton-le-Fylde 14E5
Pouso Alegre 55B3
Poŭthĭsăt 29C6
Považská Bystrica 13Q6
Povenets 22G3
Póvoa de Varzim 19B3
Povorino 23I6
Poway 49D4
Powell, Lake *resr* 46E4
Powell River 44F5
Poxoréo 53H7
Poyarkovo 30C2
Požarevac 21I2
Poza Rica 50E4
Požega 20G2
Požega 21I3
Poznań 13P4
Pozo Colorado 54E2
Pozzuoli 20F4
Prachatice 13O6
Prachuap Khiri Khan 29B6
Prado 55D2
Prague 13O5
Praia 32□
Prainha 53H4
Prairie du Chien 47I3
Prata 55A2
Prata *r.* 55A2
Prato 20D3
Pratt 46H4
Prechistoye 22I4
Preiļi 11O8
Prenzlau 13N4
Přerov 13P6
Prescott 46E5
Preševo 21I3
Presidencia Roque Sáenz Peña 54D3
Presidente Dutra 53J5
Presidente Olegário 55B2
Presidente Prudente 55A3
Presidente Venceslau 55A3
Prešov 23D6
Prespa, Lake 21I4
Presque Isle 47N2
Preston 14E5
Prestwick 16E5
Pretoria 37I3
Preveza 21I5
Pribilof Islands 44A4
Priboj 21H3
Price 46E4
Priekule 11L8
Priekuļi 11N8
Prienai 11M9
Prievidza 13Q6
Prijedor 20G2
Prijepolje 21H3
Prilep 21I4
Primorskiy Kray *admin. div.* 30D3
Primorsko-Akhtarsk 23H7
Prince Albert 44H4
Prince Charles Island 45K3
Prince Edward Island *prov.* 45L5
Prince George 44F4
Prince of Wales Island 45I2
Prince Rupert 44E4
Princess Charlotte Bay 41I2
Princeton 46C2
Princeton 48D2
Princeton 48A4
Prince William Sound *b.* 44D3
Priozersk 11Q6
Pripet *r.* 23F6
Pripet Marshes 23E6
Pristina 21I3
Privas 18G4
Privlaka 20F2
Privolzhsk 22I4
Privolzhskiy 23J6
Privolzh'ye 23K5
Priyutnoye 23I7
Prizren 21I3
Professor van Blommestein Meer *resr* 53G3
Progress 30C2
Prokhladnyy 23J8
Prokop'yevsk 24J4
Prokuplje 21I3
Proletarsk 23I7
Promissão 55A3
Propriá 53K6
Provadia 21L3
Provence *reg.* 18G5
Providence 48F2
Provideniya 25T3
Provo 46E3
Prudentópolis 55A4
Pruszków 13R4
Prut *r.* 23F7
Pryluky 23G6
Prymors'k 23H7
Przemyśl 23D6
Pskov 11P8
Pskov, Lake 11O7
Pskovskaya Oblast' *admin. div.* 11P8
Ptolemaïda 21I4
Ptuj 20F1
Pucallpa 52D5
Puchezh 22I4
Pudong 28E4
Pudozh 22H3
Pudsey 14F5
Puducherry 27G5
Puebla 50E5
Pueblo 46G4
Puente Genil 19D5
Puerto Ángel 50E5
Puerto Armuelles 51H7
Puerto Ayacucho 52E2
Puerto Baquerizo Moreno 52□
Puerto Barrios 50G5
Puerto Cabello 52E1
Puerto Cabezas 51H6
Puerto Carreño 52E2
Puerto Lempira 51H5
Puerto Limón 51H6
Puertollano 19D4
Puerto Madryn 54C6
Puerto Maldonado 52E6
Puerto Montt 54B6
Puerto Natales 54B8
Puerto Peñasco 46E5
Puerto Plata 51J5
Puerto Princesa 29D7
Puerto Rico 54E3
Puerto Rico *terr.* 51K5
Puerto Santa Cruz 54C8
Puerto Supe 52C6
Puerto Vallarta 50C4
Pugachev 23K5
Pukaki, Lake 43C7
Pukch'ŏng 31C4
Pula 20E2
Pulaski 48A4
Pul-e Khumrī 27F3
Pullman 46D2
Pune 27G5
P'ungsan 30C4
Puno 52D7
Punta Alta 54D5
Punta Arenas 54B8
Punta del Este 54F5
Punta Gorda 50G5
Puntarenas 51H6
Puntland *area* 34E3
Puri 27H5
Purus *r.* 52F4
Puryŏng 30C4
Pushchino 23H5
Pushkin 11Q7
Pushkinskiye Gory 11P8
Pustoshka 22F4
Putian 28D5
Putrajaya 29C7
Putumayo *r.* 52D4
Pwllheli 15C6
Pyaozerskiy 10Q4
Pyatigorsk 23I7
P"yatykhatky 23G6
Pye 27I5
Pyeongtaek 31B5
Pyetrykaw 23F5
Pyinmana 27I5
Pyin-U-Lwin 27I4
Pyle 15D7
Pylos 21I6
Pyŏktong 30B4
P'yŏnggang 31B5
P'yŏngsong 31B5
P'yŏngyang 31B5
Pyrenees *mts* 19H2
Pyrgos 21I6
Pyryatyn 23G6
Pyrzyce 13O4
Pytalovo 11O8

Q

Qacha's Nek 37I6
Qaidam Pendi *basin* 27I3
Qaqortoq 45N3
Qarshi 26F3
Qatar *country* 34F1
Qattara Depression 33F2
Qax 23J8
Qazax 23J8
Qazvīn 33H1
Qeqertarsuup Tunua *b.* 45M3
Qeydār 33H1
Qian'an 30B3
Qilian Shan *mts* 27I3
Qinā 33G2
Qing'an 30B3
Qingdao 28E4
Qinggang 30B3
Qingyuan 30B4
Qinhuangdao 27K3
Qin Ling *mts* 27J3
Qinzhou 27J4
Qionghai 27K5
Qiqihar 30A3
Qitaihe 30C3
Qom 26E3
Qo'qon 27G2
Qorveh 33H1
Quang Ngai 29C6
Quantock Hills 15D7
Qu'Appelle *r.* 44H4
Quartu Sant'Elena 20C5
Queanbeyan 42D5
Québec 45K5
Québec *prov.* 45K4
Queen Charlotte Sound *sea chan.* 44F4
Queen Elizabeth Islands 45H2
Queen Maud Land *reg.* 56C6
Queenscliff 42B7
Queensland *state* 42B1
Queenstown 41J8
Queenstown 43B7
Queenstown 37H6
Quelimane 35D5
Querétaro 50D4
Quetta 26F3
Quetzaltenango 50F6
Quezon City 29E6
Quibala 35B5
Quibdó 52C2
Quillabamba 52D6
Quillacollo 52E7
Quilmes 54E4
Quilpué 54B4
Quimbele 35B4
Quimper 18B3
Quimperlé 18C3
Quincy 47I4
Quincy 48F1
Quinto 19F3
Quirindi 42E3
Quirinópolis 55A2
Quitilipi 54D3
Quito 52C4
Quixadá 53K4
Quixeramobim 53K5
Qujing 27J4
Quorn 41H6
Qŭrghonteppa 27F3
Quy Nhơn 29C6
Quzhou 28D5

R

Raahe 10N4
Raasay *i.* 16C3
Raasay, Sound of *sea chan.* 16C3
Raba 29D8
Rabat 32C1
Rabaul 38F2
Rabocheostrovsk 22G2
Rach Gia 29C7
Racibórz 13Q5
Racine 47J3
Rădăuți 23E7
Radcliff 47J4
Radford 48A4
Radnevo 21K3
Radom 13R5
Radomir 21J3
Radomsko 13Q5
Radun' 11N9
Radviliškis 11M9
Radyvyliv 23E6
Rafaela 54D4
Rafsanjān 26E3
Ragusa 20F6
Raha 29E8
Rahachow 23F5
Rahimyar Khan 27G4
Raichur 27G5
Raigarh 27H4
Rainier, Mount *vol.* 46C2
Raipur 27H4
Raisio 11M6
Rajahmundry 27H5
Rajkot 27G4
Rajshahi 27H4
Rakhiv 23E6
Rakitnoye 23G6
Rakovski 21K3
Rakvere 11O7
Raleigh 47L4
Ramciel 33G4
Rame Head 15C8
Râmnicu Sărat 21L2
Râmnicu Vâlcea 21K2
Ramon' 23H6
Ramona 49D4
Ramotswa 37G3
Rampur 27G4
Ramsey 14C4
Ramsey 48D2
Ramsgate 15I7
Ramygala 11N9
Ranaghat 27H4
Rancagua 54B4
Rancharia 55A3
Ranchi 27H4
Randalstown 17F3
Randers 11G8
Råneå 10M4
Rangoon 27I5
Rangpur 27H4
Rannoch, Loch *l.* 16E4
Ranong 29B7
Rantauprapat 29B7
Rapallo 20C2
Rapid City 46G3
Rapla 11N7
Rarotonga *i.* 6
Ras Dejen *mt.* 34D2
Raseiniai 11M9
Rasht 26D3
Rasony 11P9
Rasskazovo 23I5
Ratanda 37I4
Rat Buri 29B6
Rathenow 13N4
Rathfriland 17F3
Rathlin Island 17F2
Ratnagiri 27G5
Ratne 23E6
Raton 46G4
Raul Soares 55C3
Rauma 11L6
Raurkela 27H4
Ravenna 20E2
Ravenna 48A2
Ravensburg 13L7
Rawalpindi 27G3
Rawicz 13P5
Rawlins 46F3
Rawson 54C6
Rayagada 27H5
Raychikhinsk 30C2
Rayleigh 15H7
Raymond Terrace 42E4
Raymondville 46H6
Razgrad 21L3
Razlog 21J4
Reading 15G7
Reading 48D2
Rebiana Sand Sea *des.* 33F2
Recherche, Archipelago of the *is* 40E6
Rechytsa 23F5
Recife 53L5
Recife, Cape 37G8
Recklinghausen 13K5
Reconquista 54E3
Red *r.* 47I5
Red Bank 48D2
Red Bluff 46C3
Redcar 14F4
Red Cliffs 41I6
Red Deer 44G4
Redding 46C3
Redditch 15F6
Redenção 53H5
Redlands 49D3
Red Oak 47H3
Redondo Beach 49C4
Red Sea 34D1
Red Wing 47I3
Redwood City 49A2
Ree, Lough *l.* 17E4
Reedley 49C2
Regensburg 13N6
Reggane 32D2
Reggio di Calabria 20F5
Reggio nell'Emilia 20D2
Reghin 21K1
Regina 44H4
Registro 55B4
Rehoboth 36C2
Rehoboth Bay 48D3
Reigate 15G7
Reims 18G2
Reinbek 13M4
Reindeer Lake 45H4
Relizane 19G6
Rendsburg 13L3
Renfrew 16E5
Rengo 54B4
Reni 21M2
Renmark 41I6
Rennes 18D2
Reno 46D4
Réo 32C3
Reserva 55A4
Resistencia 54E3
Reșița 21I2
Resplendor 55C2
Retalhuleu 50F6
Retford 14G5
Rethymno 21K7
Réunion *terr.* 7
Reus 19G3
Reutlingen 13L6
Revillagigedo, Islas *is* 50B5
Rewa 27H4
Rexburg 46E3
Reykjavík 10□[2]
Reynosa 46H6
Rēzekne 11O8
Rheine 13K4
Rhine *r.* 18I2
Rhine *r.* 13K5
Rhinelander 47J2
Rho 20C2
Rhode Island *state* 48F2
Rhodes 21M6
Rhodes *i.* 21M6
Rhodope Mountains 21J4
Rhône *r.* 18G5
Rhyl 14D5
Riachão 53I5
Riacho de Santana 55C1
Riacho dos Machados 55C1
Rialma 55A1
Rialto 49D3
Riau, Kepulauan *is* 27J6
Ribas do Rio Pardo 54F2
Ribble *r.* 14E5
Ribe 11F9
Ribeirão Preto 55B3
Riberalta 52E6
Rîbnița 23F7
Richards Bay 37K5
Richfield 46E4
Richland 46D2
Richmond 42E4
Richmond 49A2
Richmond 47K4
Richmond 47K4
Richmond 48C4
Rideau Lakes 47L3
Ridgecrest 49D3
Riesa 13N5
Rietavas 11L9
Rieti 20E3
Rīga 11N8
Riga, Gulf of 11M8
Riihimäki 11N6
Rijau 32D3
Rijeka 20F2
Rikuzen-takata 31F5
Rila *mts* 21J3
Rillieux-la-Pape 18G4
Rimavská Sobota 13R6
Rimini 20E2
Rimouski 45L5
Ringkøbing 11F8
Ringsted 11G9
Ringwood 42B6
Ringwood 15F8
Rio Azul 55A4
Riobamba 52C4
Rio Bonito 55C3
Rio Branco 52E6
Rio Brilhante 54F2
Rio Casca 55C3
Rio Claro 55B3
Río Cuarto 54D4
Rio de Contas 55C1
Rio de Janeiro 55C3
Rio de Janeiro *state* 55C3
Rio do Sul 55A4
Río Gallegos 54C8
Río Grande 54C8
Rio Grande 54F4
Río Grande 50D4
Rio Grande *r.* 46H6
Rio Grande City 46H6
Rio Grande do Sul *state* 55A5
Riohacha 52D1
Rioja 52C5
Río Lagartos 50G4
Rio Largo 53K5
Riom 18F4
Rio Novo 55C3
Rio Pardo de Minas 55C1
Rio Preto 55C3
Rio Rancho 46F4
Rio Verde 55A2
Rio Verde de Mato Grosso 53H7
Ripky 23F6
Ripley 15F5
Ripon 14F4
Risca 15D7
Risør 11F7
Riva del Garda 20D2
Rivas 51G6
Rivera 54E4
River Cess 32C4
Riverhead 48E2
Riverside 49D4
Riverview 45L5
Rivière-du-Loup 45L5
Rivne 23E6
Rivungo 35C5
Riyadh 34E1
Rize 23I8
Roade 15G6
Road Town 51L5
Roanne 18G3
Roanoke 48B4
Roanoke Rapids 47L4
Roaring Spring 48B2
Roatán 51G5
Robertson 36D7
Robertsport 32B4
Roberval 45K5
Robinson Ranges *hills* 40D5
Robinvale 41I6
Rocha 54F4
Rochdale 14E5
Rochefort 18D4
Rochegda 22I3
Rochester 42B6
Rochester 15H7
Rochester 47I3
Rochester 48F1
Rochester 48C1
Rochford 15H7
Rockford 47J3
Rockhampton 41K4
Rockingham 40D6
Rock Island 47I3
Rockland 48F1
Rockville 48C3
Rocky Mountains 46F3
Rodeio 55A4
Rodez 18F4
Rodniki 22I4
Roeselare 12I5
Rohnert Park 49A1
Rohrbach in Oberösterreich 13N6
Roja 11M8
Rojas 54D4
Rokiškis 11N9
Rokytne 23E6
Rolândia 55A3
Rolim de Moura 52F6
Rolla 47I4
Roman 21L1
Romania *country* 21K2
Romans-sur-Isère 18G4
Romblon 29E6
Rome 20E4
Rome 47J5
Rome 48D1
Romford 15H7
Romilly-sur-Seine 18F2
Romny 23G6
Romodanovo 23J5
Romorantin-Lanthenay 18E3
Romsey 15F8
Ronda 19D5
Rondon 54F2
Rondonópolis 53H7
Rønne 11I9
Ronneby 11I8
Roosendaal 12J5
Roquefort 18D4
Rosario 54D4
Rosário 53J4
Rosario 46D5
Rosario 50C4
Rosario 52D1
Rosário do Sul 54F4
Rosário Oeste 53G6
Roscoff 18C2
Roscommon 17D4
Roscrea 17E5
Roseau 51L5
Roseburg 46C3
Rosenberg 47H6
Rosenheim 13N7
Roseto degli Abruzzi 20F3
Rosetown 44H4
Roseville 49B1
Rosewood 42F1
Rosh Pinah 36C4
Rosignano Marittimo 20D3
Roșiori de Vede 21K2
Roskilde 11H9
Roslavl' 23G5
Rossano 20G5
Rosso 32B3
Ross-on-Wye 15E7
Rossosh' 23H6
Ross Sea 56B4
Rostock 13N3
Rostov 22H4
Rostov-na-Donu 23H7
Rosvik 10L4
Roswell 46G5
Roth 13M6
Rotherham 14F5
Rotorua 43F4
Rotterdam 12J5
Rottweil 13L6
Roubaix 18F1
Rouen 15I9
Rousay *i.* 16F1
Rovaniemi 10N3
Roven'ki 23H6
Rovereto 20D2
Rovigo 20D2
Rovinj 20E2
Rovnoye 23J6
Royal Leamington Spa 15F6
Royal Tunbridge Wells 15H7
Royal Wootton Bassett 15F7
Royston 15G6
Rozdil'na 21N1
Rtishchevo 23I5
Ruabon 15D6
Ruahine Range *mts* 43F5
Rub' al Khālī *des.* 34E2
Rubtsovsk 24J4
Ruda Śląska 13Q5
Rudnya 23F5
Rudnya 23J6
Rudnyy 26F1
Rudol'fa, Ostrov *i.* 24G1
Rufiji *r.* 35D4
Rufino 54D4
Rufisque 32B3
Rugby 15F6
Rugeley 15F6
Rügen *i.* 13N3
Ruhengeri 34C4
Ruipa 35D4
Ruiz 50C4
Rūjiena 11N8
Rukwa, Lake 35D4
Rum *i.* 16C4
Ruma 21H2
Rumāh 34E1
Rumbek 33F4
Rumphi 35D5
Runcorn 14E5
Rundu 35B5
Rusape 35D5
Ruse 21K3
Rushden 15G6
Rushworth 42B6
Russellville 47I4
Rüsselsheim 13L5
Russia *country* 24I3
Russkiy Kameshkir 23J5
Rustavi 23J8
Rustenburg 37H3
Ruston 47I5
Rutherglen 42C6
Ruthin 15D5
Rutland 48E1
Ruy Barbosa 55C1
Ruza 22H5
Ruzayevka 23J5
Ružomberok 13Q6
Rwanda *country* 34C4
Ryan, Loch *b.* 16D5
Ryazan' 23H5
Ryazhsk 23I5
Rybinsk 22H4
Rybinskoye Vodokhranilishche *resr* 22H4
Rybnik 13Q5
Rybnoye 23H5
Ryde 15F8
Rye 15H8
Ryl'sk 23G6
Ryn-Peski *des.* 23K7
Ryukyu Islands 31B8
Rzeszów 23D6
Rzhaksa 23I5
Rzhev 22G4

S

Saale *r.* 13M5
Saalfeld/Saale 13M5
Saarbrücken 13K6
Saaremaa *i.* 11M7
Saarenkylä 10N3
Saarijärvi 10N5
Saarlouis 13K6
Šabac 21H2
Sabadell 19H3
Sabae 31E6
Sabará 55C2
Sabhā 33E2
Sabinas 46G6
Sabinas Hidalgo 46G6
Sable, Cape 45L5
Sabon Kafi 32D3
Sabzevār 26E3
Săcele 21K2
Sacheon 31C6
Saco 48F1
Sacramento 55B2
Sacramento 49B1
Sacramento *r.* 49B1
Sacramento Mountains 46F5
Sada 37H7
Sádaba 19F2
Şa'dah 34E2
Sadiola 32B3
Sadovoye 23J7
Sæby 11G8
Säffle 11H7
Safford 46F5
Saffron Walden 15H6
Safi 32C1
Safonovo 23G5
Safranbolu 23G8
Saga 31C6
Sagami-nada *g.* 31E6
Sagar 27G4
Saginaw 47K3
Saginaw Bay 47K3
Sagres 19B5
Sagua la Grande 47K7
Sahagún 19D2
Sahara *des.* 32D3
Sahel *reg.* 32C3
Sahuayo 50D4
Saïda 19G6
Saijō 31D6
Saiki 31C6
Saimaa *l.* 11P6
St Agnes 15B8
St Albans 15G7
St-Amand-Montrond 18F3
St Andrews 16G4
St Ann's Bay 51I5
St Anthony 45M4
St Arnaud 42A6
St Augustine 47K6
St Austell 15C8
St-Barthélemy *i.* 51L5
St Bees Head 14D4
St Bride's Bay 15B7
St-Brieuc 18C2
St Catharines 48B1
St-Chamond 18G4
St Charles 48C3
St Charles 47I4
St Clair, Lake 47K3
St-Claude 18G3
St Clears 15C7
St Cloud 47I2
St David's Head 15B7
St-Dié-des-Vosges 18H2
St-Dizier 18G2
St Elias Mountains 44D3
Saintes 18D4
St-Étienne 18G4
St-Étienne-du-Rouvray 15I9
St-Gaudens 18E5
St George 42D2
St George 49F2
St George's 51L6
St George's Channel 17F6
St Gotthard Pass *pass* 18I3
St Helena 49A1
St Helena, Ascension and Tristan da Cunha *terr.* 6
St Helens 41J8
St Helens 14E5
St Helens, Mount *vol.* 46C2
St Helier 15E9
St Ives 15B8
St Ives 15G6
St-Jean, Lac *l.* 45K5
St-Jérôme 47M2
Saint John 45L5
St John's 51L5
St John's 45M5
St Johnsbury 47M3
St Joseph 47I4
St Just 15B8
St Kilda *is* 12C2
St Kitts and Nevis *country* 51L5
St-Laurent-du-Maroni 53H2
St Lawrence *inlet* 45L5
St Lawrence, Gulf of 45L5
St Lawrence Island 44B3
St-Lô 18D2
St-Louis 32B3
St Louis 47I4
St Lucia *country* 51L6
St Magnus Bay 16□
St-Malo 18C2
St-Malo, Golfe de *g.* 18C2
St-Marc 51J5
St-Martin *i.* 51L5
St-Médard-en-Jalles 18D4
St-Nazaire 18C3
St Neots 15G6
St-Nicolas-de-Port 18H2
St-Omer 18F1
St Paul 47I3
St Peter Port 15E9
St Petersburg 11Q7
St Petersburg 47K6
St-Pierre 45M5
St Pierre and Miquelon *terr.* 45M5
St-Quentin 18F2
St Thomas 48A1
St-Tropez 18H5
St-Vaast-la-Hougue 15F9
St Vincent, Gulf 41H7
St Vincent and the Grenadines *country* 51L6
Sakaide 31D6
Sakākā 26D4
Sakar *mts* 21L4
Sakassou 32C4
Sakata 31E5
Sakchu 31B4
Sakhalin *i.* 30F2
Sakhalinskaya Oblast' *admin. div.* 30F2
Sakhalinskiy Zaliv *b.* 30F1
Sakhile 37I4
Şäki 23J8
Saki 32D4
Šakiai 11M9
Sakon Nakhon 29C6
Sakura 31F6
Saky 23G7
Sala 11J7
Salacgrīva 11N8
Saladas 54E3
Salado *r.* 54C5
Şalālah 26E5
Salamanca 50D4
Salamanca 19D3
Salantai 11L8
Salaspils 11N8
Šalčininkai 11N9
Saldaña 19D2
Saldanha 36C7
Saldus 11M8
Sale 42C7
Salekhard 24H3
Salem 27G5
Salem 48F1
Salem 48A2
Salem 46C3
Salem 48A4
Salerno 20F4
Salerno, Golfo di *g.* 20F4
Salford 14E5
Salgótarján 13Q6
Salgueiro 53K5
Salihli 21M5
Salihorsk 11O10
Salima 35D5
Salina 46H4
Salina Cruz 50E5
Salinas 55C2
Salinas 52B4
Salinas 50D4
Salinas 49B2
Salinas *r.* 49B2
Salinópolis 53I4
Salisbury 15F7
Salisbury 48D3
Salisbury Plain 15E7
Salmi 22F3
Salmon Arm 44G4
Salmon River Mountains 46D3
Salo 11M6
Salon-de-Provence 18G5
Sal'sk 23I7
Salta 54C2
Saltash 15C8
Saltcoats 16E5
Saltillo 46G6
Salt Lake City 46E3
Salto 55B3
Salto 54E4
Salton Sea *salt l.* 49E4
Salvador 55D1
Salwah 34F1
Salween *r.* 27I4
Salzburg 13N7
Salzgitter 13M4
Samar *i.* 29E6
Samara 23K5
Samarinda 29D8
Samarqand 26F3
Sāmarrā' 33H1
Sambalpur 27H4
Sambava 35F5
Sambir 23D6
Samborombón, Bahía *b.* 54E5
Samcheok 31C5
Same 34D4
Samīrah 34E1
Şämkir 23J8
Samoa *country* 39I3
Samobor 20F2
Samoded 22I3
Samokov 21J3
Samos *i.* 21L6
Samoylovka 23I6
Sampit 29D8
Samsun 23H8
San 32C3
Şan'ā' 34E2
Sanandaj 33H1
San Angelo 46G5
San Antonio 54B4
San Antonio 46H6
San Antonio Oeste 54D6
San Benedetto del Tronto 20E3
San Bernardino 49D3
San Bernardo 54B4
San Buenaventura 46G6
San Carlos 54B5
San Carlos 52E2
San Carlos de Bariloche 54B6
San Carlos de Bolívar 54D5
San Clemente 49D4
San Cristóbal 54D4
San Cristóbal 52D2
San Cristóbal de las Casas 50F5
Sancti Spíritus 51I4
Sandakan 29D7
Sandanski 21J4
Sanday *i.* 16G1
Sandbach 15E5
Sandefjord 11G7
San Diego 49D4
Sandıklı 21N5
Sandnes 11D7
Sandnessjøen 10H3
Sandomierz 23D6
San Donà di Piave 20E2
Sandpoint 46D2
Sandur 10□1
Sandusky 47K3
Sandvika 11G7
Sandviken 11J6
San Felipe 54B4
San Felipe 46E5
San Felipe 52E1
San Fernando 54B4
San Fernando 46H7
San Fernando 29E6
San Fernando 19C5
San Fernando 51L6
San Fernando 49C3
San Fernando de Apure 52E2
Sanford 47K6
Sanford 48F1
Sanford 47L4
San Francisco 54D4
San Francisco 49A2
San Francisco Bay *inlet* 49A2
San Gabriel Mountains 49C3
Sangar 25N3
Sanger 49C2
San Giovanni in Fiore 20G5
Sangkulirang 29D7
Sangli 27G5
Sangmélima 32E4
Sangre de Cristo Range *mts* 46F4
San Ignacio 52E6
San Ignacio 52F7
San Jacinto 49D4
Sanjō 31E5
San Joaquin *r.* 49B1
San Joaquin Valley *valley* 49B2
San Jorge, Golfo de *g.* 54C7
San José 51H7
San Jose 29E6
San Jose 49B2
San Jose de Buenavista 29E6
San José de Comondú 46E6
San José del Guaviare 52D3
San José de Mayo 54E4
San Juan 54C4
San Juan 51K5
San Juan Bautista 54E3
San Juan de los Morros 52E2
San Juan Mountains 46F4
San Justo 54D4
Sankt Gallen 18I3
Sankt Pölten 13O6
Sankt Veit an der Glan 13O7
Şanlıurfa 33G1
San Lorenzo 54D4
San Lorenzo 52C3
Sanlúcar de Barrameda 19C5
San Luis 54C4
San Luis 49E4
San Luis Obispo 49B3
San Luis Potosí 50D4
San Marcos 46H6
San Marino *country* 20E3
San Marino 20E3
San Martín 54C4
San Martín de los Andes 54B6
San Mateo 49A2
San Matías, Golfo *g.* 54D6
San Miguel 50G6
San Miguel de Tucumán 54C3
Sanniquellie 32C4
Sanok 23D6
San Pedro 54D2
San-Pédro 32C4
San Pedro Channel 49C4
San Pedro de las Colonias 46G6
San Pedro de Macorís 51K5
San Pedro de Ycuamandyyú 54E2
San Pedro Sula 50G5
Sanquhar 16F5
San Rafael 54C4
San Rafael 49A2
Sanremo 20B3
San Salvador 50G6
San Salvador de Jujuy 54C2
Sansanné-Mango 32D3
San Sebastián 19F2
San Sebastián de los Reyes 19E3
San Severo 20F4
Santa Ana 50G6
Santa Ana 49D4
Santa Bárbara 55C2
Santa Bárbara 46F6
Santa Barbara 49C3
Santa Barbara Channel 49B3
Santa Bárbara d'Oeste 55B3
Santa Catalina, Gulf of 49D4
Santa Catarina *state* 55A4
Santa Clara 51I4
Santa Clara 49B2
Santa Clarita 49C3
Santa Cruz 52F7
Santa Cruz 53K5
Santa Cruz 49A2
Santa Cruz Cabrália 55D2
Santa Cruz del Sur 51I4
Santa Cruz de Tenerife 32B2
Santa Cruz do Sul 54F3
Santa Fe 54D4
Santa Fe 46F4
Santa Fé do Sul 55A3
Santa Helena 53I4
Santa Helena de Goiás 55A2
Santa Inês 53I4
Santa Maria 54F3
Santa Maria 32□
Santa Maria 49B3
Santa Maria da Vitória 55B1
Santa Maria do Suaçuí 55C2
Santa Maria Madalena 55C3
Santa Marta 52D1
Santa Monica 49C3
Santa Monica Bay 49C4
Santana 55C1
Santander 19E2
Sant'Antioco 20C5
Santa Quitéria 53J4
Santarém 53H4
Santarém 19B4
Santa Rosa 54D5
Santa Rosa 54F3
Santa Rosa 49A1
Santa Rosa de Copán 50G6
Santa Rosalía 46E6
Santa Tecla 50G6
Santa Vitória 55A2
Santee 49D4
Sant Francesc de Formentera 19G4
Santiago 54F3
Santiago 54B4
Santiago 51J5
Santiago 51H7
Santiago de Compostela 19B2
Santiago de Cuba 51I4
Santiago del Estero 54D3
Sant Jordi, Golf de *g.* 19G3
Santo Amaro 55D1
Santo Amaro de Campos 55C3
Santo Anastácio 55A3
Santo André 55B3
Santo Ângelo 54F3
Santo António 32D4
Santo Antônio da Platina 55A3
Santo Antônio de Jesus 55D1
Santo Antônio do Içá 52E4
Santo Domingo 51K5
Santorini *i.* 21K6
Santos 55B3
Santos Dumont 55C3
Santo Tomé 54E3
San Vicente 50G6
San Vicente de Cañete 52C6
Sanya 29C6
São Bernardo do Campo 55B3
São Borja 54E3
São Carlos 55B3
São Domingos 55B1
São Félix 55D1
São Félix do Araguaia 53H6
São Félix do Xingu 53H5
São Fidélis 55C3
São Francisco 55B1
São Francisco *r.* 55C1
São Francisco de Paula 55A5
São Francisco do Sul 55A4
São Gabriel 54F4
São Gonçalo 55C3
São Gonçalo do Abaeté 55B2
São Gonçalo do Sapucaí 55B3
São Gotardo 55B2
São João da Barra 55C3
São João da Boa Vista 55B3
São João da Madeira 19B3
São João da Ponte 55B1
São João del Rei 55B3
São João do Paraíso 55C1
São Joaquim 55A5
São Joaquim da Barra 55B3
São José 55A4
São José do Rio Preto 55A3
São José dos Campos 55B3
São José dos Pinhais 55A4
São Leopoldo 55A5
São Lourenço 55B3
São Luís 53J4
São Luís de Montes Belos 55A2
São Manuel 55A3
São Mateus 55D2
São Mateus do Sul 55A4
São Miguel do Araguaia 55A1
São Miguel do Tapuio 53J5
Saône *r.* 18G4
São Paulo 55B3
São Paulo *state* 55A3
São Paulo de Olivença 52E4
São Pedro da Aldeia 55C3
São Raimundo Nonato 53J5
São Romão 55B2
São Roque 55B3
São Sebastião 55B3
São Sebastião do Paraíso 55B3
São Simão 53H7
São Simão 55B3
São Tomé 32D4
São Tomé and Príncipe *country* 32D4
São Vicente 55B3
Sapanca 21N4
Sapouy 32C3
Sapozhok 23I5
Sapporo 30F4
Saqqez 33H1
Sara Buri 29C6
Sarajevo 20H3
Saraktash 24G4
Saransk 23J5
Sarapul 24G4
Sarasota 47K6
Sarata 21M1
Saratoga 49A2
Saratoga 46F3
Saratov 23J6
Saratovskoye Vodokhranilishche *resr* 23J5
Sarāvān 26F4
Saray 21L4
Sarayköy 21M6
Sardinia *i.* 20C4
Sar-e Pul 26F3
Sargodha 27G3
Sarh 33E4
Sārī 26E3
Sarıgöl 21M5
Sarina 41J4
Sariwŏn 31B5
Sarıyer 21M4
Sarkand 27G2
Şarkikaraağaç 21N5
Şarköy 21L4
Sarnen 18I3
Sarnia 47K3
Sarny 23E6
Saros Körfezi *b.* 21L4
Sarov 23I5
Sarpsborg 11G7
Sarrebourg 18H2
Sárvár 20G1
Saryarka *plain* 27G1
Sasebo 31C6
Saskatchewan *prov.* 44H4
Saskatchewan *r.* 44H4
Saskatoon 44H4
Sasolburg 37H4
Sasovo 23I5
Sassandra 32C4
Sassari 20C4
Sassnitz 13N3
Satpura Range *mts* 27G4
Satsuma-Sendai 31C7
Satu Mare 23D7
Saucillo 46F6
Sauda 11E7
Sauðárkrókur 10□[2]
Saudi Arabia *country* 26D4
Sault Sainte Marie 45J5
Sault Sainte Marie 47K2
Saumalkol' 26F1
Saumur 18D3
Saurimo 35C4
Sava *r.* 20I2
Savalou 32D4
Savannah 47K5
Savannah *r.* 47K5
Savannakhét 29C6
Savanna-la-Mar 51I5
Sävar 10L5
Savaştepe 21L5
Savona 20C2
Savonlinna 10P6
Sävsjö 11I8
Sawtell 42F3
Sawu, Laut *sea* 40E1
Saxilby 14G5
Saxmundham 15I6
Say 32D3
Sayḩūt 34F2
Saynshand 27K2
Sayreville 48D2
Scapa Flow *inlet* 16F2
Scarborough 45K5
Scarborough 51L6
Scarborough 14G4
Schaffhausen 18I3
Schärding 13N6
Schenectady 48E1
Schio 20D2
Schleswig 13L3
Schönebeck (Elbe) 13M4
Schwäbische Alb *mts* 13L7
Schwäbisch Hall 13L6
Schwandorf 13N6
Schwarzenberg/Erzgebirge 13N5
Schwaz 13M7
Schwedt/Oder 13O4
Schweinfurt 13M5
Schwerin 13M4
Schwyz 18I3
Sciacca 20E6
Scicli 20F6
Scone 42E4
Scone 16F4
Scotland *admin. div.* 16F3
Scottsbluff 46G3
Scottsboro 47J5
Scranton 48D2
Scunthorpe 14G5
Scutari, Lake 21H3
Seaford 15H8
Seamer 14G4
Searcy 47I4
Seattle 46C2
Sebba 32D3
Sebeș 21J2
Sebezh 11P8
Sebring 47K6
Sechelt 44F5
Sechura 52B5
Secunda 37I4
Secunderabad 27G5
Sedalia 47I4
Sedan 18G2
Sédrata 20B6
Šeduva 11M9
Sefare 37H2
Seferihisar 21L5
Segamat 27J6
Segezha 22G3
Ségou 32C3
Segovia 19D3
Séguéla 32C4
Seguin 46H6
Seinäjoki 10M5
Seine *r.* 15H9
Seine, Baie de *b.* 15G9

Seine, Val de valley 18F2
Sejny 11M9
Sekayu 29C8
Sek'ot'a 34M4
Sekondi 32C4
Selebi-Phikwe 35C6
Selendi 21M5
Sélibabi 32B3
Selizharovo 22G4
Selkirk 45I5
Selkirk 16G5
Selkirk Mountains 44G4
Selma 47J5
Selma 49C2
Sel'tso 23G5
Selty 22L4
Selvas reg. 52D5
Selwyn Mountains 44E3
Semarang 29D8
Semenivka 23G5
Semenov 22J4
Semey 24J7
Semikarakorsk 23I7
Semiluki 23H5
Semnān 26E3
Sena Madureira 52E5
Sendai 31F5
Senftenberg 13J5
Senegal country 32B3
Sengerema 34D4
Sengiley 23K5
Senhor do Bonfim 53J6
Senigallia 20E3
Senlis 18F2
Senqu r. 37H6
Sens 18F2
Sensuntepeque 50G6
Senta 21J2
Sento Sé 53J5
Senwabarwana 37I2
Seocheon 31B5
Seongnam 31B5
Seosan 31B5
Seoul 31B5
Sep'o 31B5
Sept-Îles 45L4
Serafimovich 23I6
Seram i. 29E8
Seram, Laut sea 29E8
Serbia country 21I3
Serdar 26E3
Serdobsk 23J5
Serekunda 32B3
Seremban 29C7
Serengeti 35D5
Sergach 22J5
Sergiyev Posad 22H4
Serik 33G1
Sernur 22K4
Serov 24H4
Serowe 37H2
Serpukhov 23H5
Serra 55C3
Serra Talhada 53K5
Serres 21J4
Serrinha 53K6
Sertãozinho 55B3
Sertolovo 11Q6
Serule 35C6
Seryshevo 30C2
Sestri Levante 20C2
Sestroretsk 11P6
Sète 18F5
Sete Lagoas 55B2
Sétif 32D1
Sète 31E6
Settle 14E4
Setúbal 19B4
Setúbal, Baía de b. 19B4
Sevan 23J8
Sevan, Lake 23J8
Sevastopol' 23G7
Sevenoaks 15H7
Severn r. 15E7
Severnaya Dvina r. 22I2
Severnaya Zemlya is 25L1
Severnyy 24H3
Severodvinsk 22I2
Severomorsk 10R2
Severo-Yeniseyskiy 24K3
Severskaya 23H7
Sevilla 52C3
Seville 19D5
Seward 44D3
Seychelles country 7
Seymchan 25Q3
Seymour 47J6
Sfântu Gheorghe 21K2
Sfax 20D7
Shaftesbury 15E7
Shahdol 27H4
Shahr-e Kord 26E3
Shahrisabz 26F3
Shakhovskaya 22G4
Shakhty 23I7
Shakhun'ya 22J4
Shalakusha 22I3
Shali 23J8
Shalkar 26E2
Shamrock 46G4
Shandong Bandao pen. 28E4
Shanghai 28E4
Shangzhi 30B3
Shanhe 30B3
Shannon est. 17D5
Shannon r. 17D5
Shannon, Mouth of the 17C5
Shantou 28D5
Shaoyang 27K4
Shapinsay i. 16G1
Shaqrā' 33H2
Sharjah 26E2
Sharkawshchyna 11O9
Shark Bay 40C5
Sharon 48A2
Shar'ya 22J4
Shashemenē 34D3
Shatki 23J5
Shatsk 23J5
Shatura 23H5
Shawano 47J3
Shawnee 47H4
Shchekino 23H5
Shchigry 23H5
Shchor 23H6
Shchuch'yn 11N10
Shebekino 23H6
Shebelē Wenz, Wabē r. 34E3
Sheboygan 47J3
Shebunino 30F3
Sheerness 15H7
Sheffield 14F5
Shelburne Bay 41I2
Shelbyville 47J4
Shenandoah Mountains 48B3
Shendam 32D4
Shenkursk 22I3
Shentala 23K5

Shenyang 30A4
Shepetivka 23E6
Shepparton 42B6
Sheppey, Isle of i. 15H7
Sherbrooke 45K5
Sheridan 46F3
Sherman 47H5
's-Hertogenbosch 12J5
Sherwood Forest reg. 15F5
Shetland Islands 16¹
Shetpe 26E2
Sheyenne r. 46H2
Shibata 31E5
Shibirghān 26F3
Shiel, Loch l. 16D4
Shihezi 27J2
Shijiazhuang 27K3
Shikarpur 27G4
Shikoku i. 31D6
Shilda 23L5
Shiliguri 27H4
Shillong 27I4
Shilovo 23J5
Shimada 31E6
Shimla 27H3
Shimoga 27G5
Shimonoseki 31C6
Shin, Loch l. 16E2
Shinnston 48A3
Shiogama 31F5
Shīrāz 26E4
Shivpuri 27G4
Shiyan 27K3
Shizhong 27K3
Shizuishan 27J3
Shizuoka 31E6
Shklow 23F5
Shkodër 21H3
Shōbara 31D6
Shoshong 37H2
Shostka 23G6
Shpola 23F6
Shreveport 47I5
Shrewsbury 15E6
Shuangcheng 30B3
Shuangliao 30A4
Shuangyang 30B4
Shuangyashan 30C3
Shubarkuduk 26E2
Shulan 30B3
Shumen 21L3
Shumerlya 22J5
Shumilina 23F5
Shumyachi 23G5
Shūnan 31C6
Shuya 22J4
Shuya 22J5
Shyrokent 27F2
Shyrokye 23G7
Šiauliai 11M9
Sibasa 37J2
Šibenik 20F3
Siberia reg. 25M3
Sibi 34B4
Sibiti 34B4
Sibiu 21K2
Sibolga 27I6
Sibu 34B3
Sibut 34B3
Sichuan Pendi basin 27J3
Sicilian Channel 20E6
Sicily i. 20F5
Sicuani 52D6
Sidi Aïssa 19I6
Sidi Ali 19G5
Sidi Bel Abbès 19F6
Sidi Bouzid 20C7
Sidi Ifni 32B2
Sidi Kacem 32C1
Sidley Hills 46I4
Sidmouth 15D8
Sidney 46G2
Sidney 47K3
Siedlce 11M10
Siegen 13L5
Siena 20D3
Sieradz 13J5
Sierra Grande 54C4
Sierra Leone country 32B4
Sierra Madre Mountains 49B3
Sierra Vista 46F5
Sierre 18H3
Sig 19F6
Sighetu Marmației 23D7
Sighișoara 21K1
Sigli 27I6
Siguiri 32C3
Sihanoukville 29C6
Siilinjärvi 10O5
Sikar 27G4
Sikasso 32C3
Sikeston 47J4
Sikhote-Alin' mts 30D4
Silale 11M9
Silchar 27I4
Şile 21M4
Silesia reg. 13P5
Siliana 20C6
Silifke 33G1
Silistra 21L2
Silivri 21M4
Siljan l. 11I6
Silkeborg 11F8
Silkwood 41I8
Sillamäe 11O7
Silsbee 47I5
Silute 11L9
Silvânia 55A2
Silver City 46F5
Silver Spring 48C3
Simav 21L5
Simcoe 48A2
Simcoe, Lake 48K5
Simeonovgrad 21L3
Simeulue i. 27I6
Simferopol' 23G7
Simi Valley 49C3
Simleu Silvaniei 21J1
Šimplicio Mendes 53J5
Simpson Desert 41H4
Simrishamn 11I9
Simulcsejo 52C2
Sindelfingen 13L6
Sındırgı 21M5
Sindou 32C3
Sines 19B4
Singapore country 29C7
Singapore 29C7
Singkawang 29C7
Singleton 42E4
Siniscola 20C4
Sinjai 29E8
Sinnamary 53J2
Sinop 23H7
Sinp'o 31C4
Sinsang 31B5
Sint Eustatius i. 51L5
Sint Maarten terr. 51L5
Sint-Niklaas 12J5
Sintra 19B4
Sinŭiju 31A4
Siófok 20H1
Sioux City 47H3
Sioux Falls 47H3
Siping 30B4

Sir Edward Pellew Group is 41H3
Sırnak 26C2
Sirsa 27G3
Sirte 33E1
Sirte, Gulf of 33E1
Sisak 20G2
Sitapur 27H4
Siteki 37J3
Sítio do Mato 55C1
Sitka 44E4
Sittingbourne 15H7
Sittwe 27I4
Sivas 26C3
Sivaslı 21M5
Sivrihisar 21N5
Siwah, Wāḩāt oasis 33F2
Siyabuswa 37I3
Sjenica 21I3
Sjöbo 11I9
Skadovs'k 23O1
Skagafjörður inlet 10C²
Skagen 11G7
Skagerrak strait 11F8
Skanderborg 11F8
Skara 11H7
Skarżysko-Kamienna 13R5
Skawina 13Q6
Skegness 14H5
Skellefteå 10L4
Skellefteälven 10L4
Skelmersdale 14E5
Skien 11F7
Skierniewice 13R5
Skikda 20B6
Skipton 14E5
Skive 11F8
Skjern 11F9
Skopin 23H5
Skopje 21I4
Skövde 11H7
Skovorodino 30A1
Skowhegan 47N3
Skuodas 11L8
Skurup 11H9
Skutskär 11J6
Skvyra 23F6
Skye i. 16C3
Skyros 21K5
Slagelse 11G9
Slantsy 11P7
Slatina 21K2
Slave Coast 32D4
Slavgorod 24I4
Slavonski Brod 20H2
Slavuta 23E6
Slavutych 23F6
Slavyanka 30C4
Slavyansk-na-Kubani 23H7
Sławno 13P3
Sleaford 15G5
Sleat, Sound of sea chan. 16D3
Slieve Bloom Mountains hills 17E5
Slieve Donard hill 17G3
Sligo 17D3
Sligo Bay 17D3
Slippery Rock 48A2
Sliven 21L3
Slobodskoy 22K4
Slobozia 21L2
Slonim 11N10
Slough 15G7
Slovakia country 13Q6
Slovenia country 20F2
Slovenj Gradec 20F1
Slov"yans'k 23I7
Słupsk 13P3
Slutsk 11O10
Slyudyanka 27J1
Smallwood Reservoir 45L4
Smalyavichy 11P9
Smarhon' 11O9
Smederevo 21I2
Smederevska Palanka 21I2
Smidovich 30D2
Smila 23F6
Smithton 41J8
Smithton 47J4
Smolensk 23G5
Smolyan 21K4
Snake r. 46D2
Snake River Plain 46E3
Snares Islands 39G6
Sneek 12J4
Snettisham 15H6
Snihurivka 23G7
Snizort, Loch b. 16C3
Snowdon mt. 15C5
Snowy r. 42D6
Snowy Mountains 42C6
Snyder 46G5
Soalala 35E5
Soanierana-Ivongo 35E5
Sobinka 22I5
Sobral 53J4
Sochaczew 13R4
Sochi 23H8
Society Islands 6
Socorro 52C2
Socorro 46F5
Socotra i. 26D5
Sodankylä 10O3
Söderhamn 11J6
Söderköping 11J7
Södertälje 11J7
Sodo 34D3
Sofia 21J3
Sofiya 34E2
Søgne 11E7
Sognefjorden inlet 11D6
Söğüt 21N4
Soissons 18F2
Sokal' 23E6
Söke 21L6
Sokhumi 23I8
Sokodé 32D4
Sokol 22I4
Sokoto 32D3
Sokyryany 23E6
Solana Beach 49D3
Solapur 27G5
Soledad 54F3
Solenoye 23I7
Solginskiy 22I3
Soligalich 22I4
Solihull 15F6
Solikamsk 24G4
Sollefteå 10J5
Solnechnogorsk 22H4
Solnechnyy 30E2
Solomon Islands country 38G3
Solomon Sea 38F2
Solothurn 18H3
Solov'yevsk 30B1
Sol'tsy 22F4
Sölvesborg 11I8
Solway Firth est. 16F6
Solwezi 35C5
Somalia country 34E3

Somaliland disp. terr. 34E3
Sombor 21H2
Somero 11N6
Somerset 47K4
Somerset Island 45I2
Somerset West 36D8
Somersworth 48F1
Somerville 48E2
Songbu 30C4
Sŏnch'ŏn 31B5
Sønderborg 11F9
Sondershausen 13M5
Sondrio 20C1
Songea 35D5
Songgianghe 30B4
Songkhla 29C7
Songnim 31B5
Songo 35B5
Songo 35D5
Songyuan 30B3
Sonkel', Ozero l. 26F2
Son La 28C5
Sonoran Desert 49F4
Sonqor 33H1
Sonsonate 50G6
Sopot 21K3
Sopot 13Q3
Sopron 20G1
Sop's Arm 45S2
Söråker 10J5
Sorel 47M2
Soria 19F3
Soroca 23F6
Sorocaba 55B3
Sorong 29F8
Soroti 34D3
Sorsogon 29E6
Sortavala 10Q6
Soshanguve 37I3
Sosnogorsk 22L3
Sosnovka 23I5
Sosnovyy Bor 11P7
Sosnowiec 13Q5
Sotteville-lès-Rouen 15I9
Soubré 32C4
Soufrière 51L6
Soufrière vol. 51L6
Sougueur 19G6
Souk Ahras 20B6
Souk el Arbaâ du Rharb 32C1
Soulac-sur-Mer 18D4
Sour el Ghozlane 19H5
Sousa 53K5
Sousse 20D7
Souterraine 18E3
Southampton 15F8
South Anston 14F5
South Australia state 40G6
South Bend 47J3
South Carolina state 47K5
South China Sea 29D6
South Dakota state 46G3
South Downs hills 15G8
South East admin. dist. 37G3
Southend-on-Sea 15H7
Southern admin. dist. 36G3
Southern Alps mts 43C6
Southern Uplands hills 16E5
South Georgia i. 54I8
South Georgia and South Sandwich Islands terr. 6
South Harris pen. 16B3
South Island 43D7
South Korea country 31B5
South Lake Tahoe 49B1
Southminster 15H7
South Mountains hills 48C3
Southport 14D5
South Ronaldsay i. 16G2
South San Francisco 49A2
South Shields 14F3
South Sudan country 33G4
South Taranaki Bight b. 43E4
South Uist i. 16B3
South West Cape 43A8
Southwold 15I6
Soutpansberg mts 37I2
Sovetsk 11L9
Sovetsk 22K4
Sovetskaya Gavan' 30F2
Sovetskiy 24H3
Sovetskiy 22K4
Sovyets'kyy 23G7
Soweto 37H3
Spain country 19E3
Spalding 15G6
Spanish Town 51I5
Sparks 49B1
Spartanburg 47K5
Sparti 21J6
Spas-Demensk 23G5
Spas-Klepiki 23I5
Spassk-Dal'niy 30D3
Spassk-Ryazanskiy 23I5
Spencer 47H3
Spencer Gulf est. 41H6
Sperrin Mountains hills 17E3
Spey r. 16F3
Speyer 13L6
Spilsby 14H5
Spirovo 22H4
Spišská Nová Ves 23J6
Spittal an der Drau 13N7
Split 20G3
Spokane 46D2
Spoleto 20E4
Spratly Islands 29D6
Springbok 36C6
Springdale 45M5
Springer 46G4
Springfield 47I4
Springfield 47J4
Springfield 47N3
Spring Hill 47K6
Spring Valley 48D2
Srebrenica 21H2
Sredets 21L3
Srednyy Khrebet mts 25Q4
Sredna Gora mts 21J3
Srednyaya Akhtuba 23J6
Sretensk 27K1
Sri Aman 29C7

Sri Jayewardenepura Kotte 27G6
Sri Lanka country 27H6
Srinagar 27G3
Srivardhan 27G5
Stade i. 16C4
Stadskanaal 13K4
Staffa i. 16C4
Stafford 15E6
Staines-upon-Thames 15G7
Stakhanov 23H6
Stalbridge 15E8
Stalham 15I6
Stalowa Wola 13D6
Stamford 15G6
Stamford 48E2
Standerton 37I4
Stanley 54E8
Stanley 14F4
Stannington 14F3
Stanovoy Nagor'ye mts 25M4
Stanovoy Khrebet mts 25N4
Stanthorpe 42E2
Stanton 15H6
Starachowice 13R5
Staraya Russa 22F4
Stara Zagora 21K3
Stargard Szczeciński 13O4
Staritsa 22G4
Starkville 47J5
Starobil's'k 23H6
Starogard Gdański 13Q4
Starokostyantyniv 23E6
Starominskaya 23H7
Staroshcherbinovskaya 23H7
Staryya Darohi 23F5
Staryy Oskol 23H6
State College 48C2
Statesboro 47K5
Staunton 48B3
Stavanger 11D7
Staveley 14F5
Stavropol' 23I7
Stavropol'skaya Vozvyshennost' hills 23I7
Steamboat Springs 46F3
Steinkjer 10G4
Steinkopf 36C7
Stellenbosch 36D7
Stendal 13M4
Stenungsund 11G7
Stephenville 45H5
Stepnoye 23J6
Sterkfontein Dam resr 37I5
Sterling 46G3
Sterlitamak 24G4
Steubenville 48A2
Stevenage 15G7
Stewart Island 43A8
Steynsburg 37H6
Steyr 13O6
Stikine Plateau 44E4
Stillwater 47H4
Stilton 15G6
Štip 21J4
Stirling 16F4
Stjørdalshalsen 10G5
Stockholm 11K7
Stockport 14E5
Stockton 49B2
Stockton-on-Tees 14F4
Stoke-on-Trent 15E5
Stokesley 14F4
Stolac 20G3
Stolin 11O11
Stone 15E6
Stonehaven 16G4
Storm Lake 47H3
Stornoway 16C2
Storozhynets' 23E6
Storrs 48E2
Storuman l. 10J4
Storvik 11J6
Stour r. 15F8
Stourbridge 15E6
Stourport-on-Severn 15E6
Stowbtsy 11O10
Stowmarket 15H6
Strabane 17E3
Strakonice 13N6
Stralsund 13N3
Strand 36D8
Strangford Lough inlet 17G3
Stranraer 16D6
Strasbourg 18H2
Stratford 42C6
Stratford 48A1
Stratford-upon-Avon 15F6
Strathspey valley 16F3
Stratton 15C8
Straubing 13N6
Streaky Bay 40G6
Street 15E7
Strehaia 21J2
Strenči 11N8
Stromboli, Isola i. 20F5
Strømstad 11G7
Stronsay i. 16G1
Stroud 15E7
Struer 11F8
Struga 21I4
Strugi-Krasnyye 11P7
Struma r. 21J4
Strumica 21J4
Struthers 48A2
Strydenburg 36F5
Stryy 23D6
Stupino 23H5
Sturgis 46G3
Sturt Creek watercourse 39E9
Sturt Plain 40G3
Sturt Stony Desert 41I5
Stutterheim 37H7
Stuttgart 13L6
Stuttgart 47I5
Suakin 33H3
Subotica 21H1
Suceava 23F7
Sucre 52E7
Sudak 23G7
Sudan country 33F3
Sudbury 45J5
Sudbury 15H6
Sudeten mts 13P5
Sudislavl' 22I4
Sudogda 22I5
Sueca 19F4
Suez 33G2
Suez, Gulf of 33G2
Suez Canal 33G1
Suffolk 47L4
Sūhāj 33G2
Şuḩār 26E4
Sühbaatar 27J1
Suhl 13M5
Suhut 21N5
Suibin 30C3
Suifenhe 30C4
Suihua 30B3
Suileng 30B3
Suir r. 17E5
Suizhou 27K3
Sukabumi 29C8

Sukagawa 31F5
Sukchŏn 31B5
Sukhinichi 23G5
Sukhona r. 22I3
Sukkur 27F4
Sulaiman Range mts 27F3
Sullana 52B4
Sulmona 20E3
Sulphur Springs 47H5
Sulu Archipelago is 29E7
Sulu Sea 29D7
Sumatra i. 29B7
Sumba i. 40D1
Sumba, Selat sea chan. 29D8
Sumbawa i. 40D1
Sumbawanga 35D4
Sumburgh Head 16²
Šumperk 13P6
Sumqayıt 26D2
Sumter 47K5
Sumy 23G6
Sunbury 42B6
Sunbury 48C2
Suncheon 31B6
Sunderland 14F4
Sundsvall 10J5
Sunndalsøra 10F5
Sunnyvale 49A2
Suntar 25M3
Sunyani 32C4
Suoyarvi 22G3
Superior 46H3
Superior 47I2
Superior, Lake 47J2
Şūr 26E4
Surabaya 29D8
Surakarta 29D8
Surat 27G4
Surat Thani 29B7
Surazh 23G5
Surdulica 21J3
Surgut 24I3
Surigao 29E7
Surin 29C6
Suriname country 53G3
Sürmene 23I8
Surovikino 23I6
Surskoye 23J5
Surtsey i. 10³
Susanville 46C3
Susaman 25P3
Susanville 49B1
Susurluk 21M5
Sutherland 42E5
Sutter 49B1
Sutton 15H6
Sutton Coldfield 15F6
Sutton in Ashfield 15F5
Suva 39H3
Suvorov 23H5
Suwa 31E5
Suwałki 11M9
Suwon 31B5
Suzaka 31E5
Suzhou 28E4
Suzuka 31E6
Svalbard terr. 24C2
Svatove 23I6
Svecha 22J4
Svenčionys 11O9
Svendborg 11G9
Svetlaya 30E3
Sveti Nikole 21I4
Svetlogorsk 11L9
Svetlograd 23I7
Svetlyy 11L9
Svetlyy Yar 23J6
Svilengrad 21L4
Svishtov 21K3
Svitavy 13P6
Svitlovods'k 23G6
Svobodnyy 30C2
Svyetlahorsk 23F5
Swadlincote 15F6
Swains Island atoll 39I3
Swakopmund 36B2
Swale r. 14F4
Swanage 15F8
Swan Hill 42A5
Swanley 15H7
Swansea Bay 15D7
Swaziland country see Eswatini
Sweden country 10I5
Sweetwater 46G5
Swellendam 36E8
Świdnica 13P5
Świdwin 13O4
Świebodzin 13O4
Świecie 13Q4
Swindon 15F7
Świnoujście 13O4
Switzerland country 18I3
Syców 13Q5
Sydney 42E4
Syeverodonets'k 23H6
Syktyvkar 22K3
Sylhet 27I4
Synel'nykove 23G6
Syracuse 20F6
Syracuse 48C1
Syrdar'ya r. 26F2
Syria country 33G1
Syrian Desert 33G1
Syumsi 22J4
Syzran' 23J5
Szczecin 13O4
Szczecinek 13P4
Szczytno 13R4
Szeged 21I1
Székesfehérvár 20H1
Szekszárd 20H1
Szentes 21I1
Szentgotthárd 20G1
Szigetvár 20G1
Szolnok 21I1
Szombathely 20G1

T

Tābah 34E1
Tabatinga 52E4
Tablógm 32D4
Tabora 35D4
Tabou 32C4
Tabrīz 26D3
Tabūk 26C3
Tāby 11K7
Tacheng 27I2
Tachov 13N6
Tacloban 29E6
Tacna 52D7
Tacoma 52D7
Tacuarembó 54F4
Tadcaster 14F5
Tademaït, Plateau du 32D2
Tadjourah 34E2
Tadmur 33G1
Taegu 31B5
Tafalla 19E3
Tafí Viejo 54C3
Tafresh 33H1
Taganrog 23H7

Taganrog, Gulf of 23H7
Taguranga 43F3
Tahiti i. 6
Tahlequah 47H4
Tahoua 32D3
Tai'an 27K3
Taidong 28E5
Tailai 30A3
Tainan 28E5
Taiobeiras 55C1
Taipei 28E5
Taiping 29C7
Taiwan country 28E5
Taiwan Strait strait 28D5
Ta'izz 34E2
Tajikistan country 27G3
Tak 29B6
Takāb 33H1
Takahashi 31D6
Takamatsu 31E6
Takaoka 31E5
Takapuna 43E3
Takayama 31E5
Takhemaret 19G6
Takum 32D4
Talachyn 23F5
Talagang 27G3
Talara 52B4
Talavera de la Reina 19D4
Talca 54B5
Talcahuano 54B5
Taldom 22H4
Taldykorgan 27H2
Tallahassee 47K5
Tallinn 11N7
Tallulah 47I5
Tal'ne 23F6
Talovaya 23I6
Talsi 11M8
Tamala 23I5
Tamale 32C4
Tamano 31D6
Tamanrasset 32D2
Tambacounda 32B3
Tambov 23I5
Tambovka 30C2
Tampa 47K6
Tampere 11M6
Tampico 50E4
Tamworth 42E3
Tamworth 15F6
Tana r. 34E4
Tana, Lake 34D2
Tanabe 31D6
Tanabi 55A3
Tanami Desert 40G3
Tanch'ŏn 31C4
Tanda 32C4
Tăndărei 21L2
Tandil 54E5
Tandragee 17F3
Tanezrouft reg. 32C2
Tanga 35E4
Tanganyika, Lake 35C4
Tangará 55A4
Tanggula Shan mts 27H3
Tangier 19D6
Tangra Yumco salt l. 27H3
Tangshan 27K3
Tangyuan 30C3
Tanhua 55C1
Tanjay 29E7
Tanjungredeb 29D7
Tanjungselor 29D7
Tanout 32D3
Ţanţā 33G1
Tan-Tan 32B2
Tanzania country 35D4
Taonan 30A3
Taourirt 32C1
Tapachula 50F6
Tapajós r. 53G4
Tapauá 52E5
Taperoá 55D1
Taquara 55A5
Taquari r. 55A1
Tarakan 29D7
Taraklı 21N4
Taranto 20G4
Taranto, Golfo di g. 20G4
Tarapoto 52C5
Tararovskiy 23I6
Tarauacá 52D5
Taraz 27G2
Tarbert 16E5
Tarbes 18E5
Tarfaya 32B2
Targovishte 21L3
Târgoviște 21K2
Targuist 19D6
Târgu Jiu 21J2
Târgu Mureş 21K1
Târgu Neamţ 21L1
Târgu Secuiesc 21L1
Tarif 34F1
Tarija 52E7
Tarim Basin 27H3
Tarime 34D4
Tarko-Sale 24I3
Tarkwa 32C4
Tarlac 29E6
Tarma 52C6
Tărnăveni 21K1
Tarnobrzeg 23D6
Tarnogskiy Gorodok 22I3
Tarnów 23D6
Tarnowskie Góry 13Q5
Taroudannt 32C1
Tarragona 19G3
Tarrafal 32□
Tarsus 33G1
Tartagal 52F8
Tartās 26C3
Tărtăr 23J8
Tartu 11O7
Tashir 23J8
Taşköprü 23H7
Tasman Bay 43D5
Tasmania state 41J8
Tasman Sea 38H6
Tasova 23H7
Tata 20H1
Tatabánya 20H1
Tatarbunary 21M2
Tatarsk 24I4
Tateyama 31E6
Tatinskiy 34E2
Tatra Mountains 13Q6
Tatsinskaya 23I6
Tatuí 55B3
Tatvan 26D3
Tauá 53J5
Taubaté 55B3
Taunggyi 27I5
Taungup 27I5
Taunton 15D7
Taupo 43F4
Taupo, Lake 43E4

Taurag 11M9
Tauranga 43F3
Taurus Mountains 26C3
Tavas 21M6
Tavira 19C5
Tavistock 15C8
Tavşanlı 21M5
Tavua 39H3
Tay r. 16F4
Tay, Firth of est. 16F4
Taylor 47I5
Taymā' 26C4
Taymyr, Ozero l. 25L2
Taymyr Peninsula 24J3
Tây Ninh 29C6
Taz r. 24J3
Tazovskiy 24I3
Tbilisi 23J8
Tchamba 32D4
Tchibanga 34B4
Tczew 13Q3
Te Anau, Lake 43A7
Teapa 50F5
Tébarat 32D3
Tébessa 20C7
Tébourba 20C6
Tecate 49D4
Tecomán 50D5
Tecpan 50D5
Tecuala 50C4
Tecuci 21L1
Tees r. 14F4
Tefenni 21M6
Tegucigalpa 51G6
Tehrān 26E3
Tehuacán 50E5
Tehuantepec, Gulf of 50F5
Teignmouth 15D8
Teixeiras 55C3
Teixeira Soares 55A4
Tejen 26F3
Tejen r. 26F3
Tekax 50G4
Tekirdağ 21L4
Telangana state 27H5
Telêmaco Borba 55A4
Telford 15E6
Tellisheim 13283
Tel'novskiy 30F2
Telšiai 11M9
Tembisa 37I3
Temecula 49D4
Temir 26F2
Temirtau 27G1
Temnikov 23I5
Temora 42C4
Temple 47H5
Temryuk 23H7
Temuco 54B5
Tena 52C4
Tenali 27H5
Tenby 15C7
Tendō 31F5
Ténéré du Tafassâsset des. 32D2
Tenerife i. 32B2
Ténés 19G5
Tengréla 32C3
Tekeli 25P2
Ten-kodogo 32C3
Tennant Creek 40G3
Tennessee r. 47J4
Tennessee state 47J4
Tenosique 50F5
Tenterfield 42E2
Teodoro Sampaio 55A3
Teófilo Otoni 55C2
Tepatitlán 50D4
Tepic 50D4
Teplice 13N5
Teploye 23H5
Tequila 50D4
Téra 32D3
Teramo 20E3
Terang 42A7
Terebovlya 23E6
Teresina 53J5
Teresópolis 55C3
Teribërka 10R2
Termas de Río Hondo 54D3
Termez 26F3
Termoli 20F4
Ternate 29E7
Terneuzen 12I5
Terni 20E4
Ternopil' 23E6
Terra Bella 49C3
Terrace 44F4
Terre Haute 47J4
Teruel 19E3
Teseney 34D2
Tessaoua 32D3
Tete 35D5
Tetiyiv 23F6
Tétouan 19D6
Tetovo 21I3
Tetyushi 23K5
Tewantin 41K5
Texarkana 47I5
Texas state 46H5
Teyateyaneng 37H5
Teykovo 22I4
Thaba Nchu 37H5
Thaba-Tseka 37I5
Thabong 37H4
Thai Binh 29C5
Thailand country 29C6
Thailand, Gulf of 28C5
Thai Nguyên 28C5
Thakhek 29C6
Thamaga 37G3
Thames r. 15H7
Thames r. 15H7
Thandwe 27I5
Thanet, Isle of pen. 15I7
Thanh Hoa 29C6
Thanjavur 27G5
Thar Desert 27F4
Thaton 27I5
Thayetmyo 27I5
The Bahamas country 51I4
The Dalles 46C2
The Entrance 42E4
The Fens reg. 15G6
The Gambia country 32B3
The Gulf 26A4
The Hague 12J4
The Minch sea chan. 16C2
Theniet el Had 19H6
The North Sound sea chan. 16G1
Thermaïkos Kolpos g. 21J4
The Solent strait 15F8
Thessaloniki 21J4
Thetford 15H6
Thetford Mines 47M2
The Valley 51L5

Tsaratanana, Massif du mts 35E5
Tsetserleg 27J2
Tshabong 36F4
Tshela 35C4
Tshikapa 35C4
Tsimlyansk 23I7
Tsimlyanskoye Vodokhranilishche resr 23I7
Tsiroanomandidy 35E5
Tsivil'sk 22J4
Tskhinvali 23I8
Tsna r. 23I5
Ts'nori 23J8
Ts'q'alt'ubo 23I8
Tsu 31E6
Tsumeb 35B5
Tsuruga 31E6
Tsuruoka 31E5
Tswelelang 37G4
Tsyelyakhany 11N10
Tsyurupyns'k 23O1
Tuamotu Islands 6
Tuapse 23H7
Tubarão 55A5
Tübingen 13L6
Tubmanburg 32B4
Tubuai 33□
Tubruq 33E1
Tucano 53K6
Tucson 46G5
Tucumcari 46G4
Tucupita 52F2
Tucuruí 53I4
Tucuruí, Represa de resr 53I4
Tudela 19F3
Tudun Wada 32D3
Tuguegarao 29E6
Tukums 11M8
Tukuyu 35D4
Tula 23H5
Tulaghi 41M1
Tulancingo 50E4
Tulare 49C2
Tulcán 52C3
Tulcea 21M2
Tulihe 30A2
Tullamore 17E4
Tulle 18E4
Tullow 17F5
Tulsa 43H4
Tulsa 47H4
Tuluá 52C3
Tumaco 52C3
Tumahole 37H4
Tumba 11J7
Tumbarumba 42D5
Tumbes 52B4
Tumby Bay 41H6
Tumen 30C4
Tumu 32C3
Tumucumaque, Serra hills 53G3
Tunceli 26C3
Tuncurry 42F4
Tunduru 35D5
Tungor 30F1
Tunis 20D6
Tunis, Golfe de g. 20D6
Tunisia country 32D1
Tunja 52D2
Tupã 55A3
Tupelo 47J5
Tupiza 52E8
Tupungato, Cerro mt. 54C4
Tura 25L3
Turan Lowland 26F2
Turbo 52C2
Turda 21J1
Turgutlu 21L5
Turin 20B2
Turkana, Lake salt l. 34D3
Turkey country 26C3
Türkistan 26F2
Türkmenabat 26F3
Türkmenbaşy 26E2
Turkmenistan country 26E2
Turks and Caicos Islands terr. 51J4
Turku 11M6
Turkwel watercourse 34D3
Turnhout 12J5
Turnu Măgurele 21K3
Turpan 27H2
Turriff 16G3
Tuscaloosa 47J5
Tuscarora Mountains hills 48C2
Tuskegee 47J5
Tussey Mountains hills 48B2
Tutayev 22H4
Tuticorin 27G6
Tuttlingen 13L7
Tutubu 35D4
Tuvalu country 39H2
Tuwayq, Jabal mts 34E1
Tuxpan 50E4
Tuxpan 50C4
Tuxtla Gutiérrez 50F5
Tuy Hoa 29C6
Tuz, Lake salt l. 26C3
Tuzi, Lake salt l. 26C3
Tuz Khurmātū 33H1
Tuzla 21H2
Tver' 22G4
Tweed r. 16G5
Tweed Heads 42F2
Twentynine Palms 49D3
Twin Falls 46E3
Twizel 43C7
Tyler 47H5
Tymovskoye 30F2
Tynda 30B2
Tynemouth 14F3
Tywyn 15C6

U

Uaua 53K5
Ubá 55C3
Ubai 55B2
Ubaitaba 55D1
Ubangi r. 34B4
Ube 31C6
Úbeda 19D4
Uberaba 55B2
Uberlândia 55A2
Ubon Ratchathani 29C6
Ucar 23J8
Uçarı r. 52D4
Uchiura-wan b. 30F4
Uckfield 15H8
Udaipur 27G4
Uddevalla 11G7
Udimskiy 22J3